geography@university

making the most of your geography degree and courses

Gordon Clark and Terry Wareham

SAGE Publications
London • Thousand Oaks • New Delhi

First published 2003

SAGE Publications Ltd
6 Bonhill Street
London EC2A 4PU

SAGE Publications Inc.
2455 Teller Road
Thousand Oaks, California 91320

SAGE Publications India Pvt Ltd
32, M-Block Market
Greater Kailash – 1
New Delhi 110 048

British Library Cataloguing in Publication Data
A catalogue record for this book is available from the British Library

ISBN 0 7619 4025 1
ISBN 0 7619 4026 X (pbk)

Library of Congress Number 2002 105444

Typeset by Keystroke, Jacaranda Lodge, Wolverhampton.
Printed in Great Britain by Athenaeum Press, Gateshead

CONTENTS

PREFACE

We think it will be worthwhile to read this Guide because it:

- tells you how geography degrees work, so you will know what to expect;

- gives you advice on making the most of your geography teaching and how to learn geography successfully;

- will help you do better in examinations, essays and the other ways in which your geography courses will be assessed;

- focuses on how geography will give you new skills and help you develop as a person, colleague and citizen;

- will prepare you for a career and learning long after university.

We hope you find this a useful Guide.

Gordon Clark and Terry Wareham

ACKNOWLEDGEMENTS

The authors gratefully acknowledge:

- the assistance of the Geography Discipline Network (GDN) under whose auspices, and with funding from the former Department for Education and Employment, an earlier version of this book was produced;
- the advice and comments from geography colleagues and members of the Advisory and Graduate Panels of the GDN, and from undergraduates in the Department of Geography at Lancaster University,
- the cartoons in Sections 4.2, 4.3, 5.4 and 5.8 drawn by Dr Peter Vujakovic (Christchurch College, Canterbury, Kent) and the cartoon in Section 4.12 drawn by Dr Mike Jeffries (University of Northumbria), which are reproduced by kind permission of the artists and the editors and publisher (Carfax) of the *Journal of Geography in Higher Education* in which they first appeared;
- the cartoons in Sections 4.1 and 4.5 devised by Dr Gordon Clark and drawn by Christopher Beacock (Lancaster University);
- the quotations from geographers supplied by Dr Lorraine Craig (Royal Geographical Society with the Institute of British Geographers, London);
- the assistance of Dr Iain Hay (Flinders University, Adelaide, Australia) and Professor Richard Le Heron (University of Auckland, New Zealand) for the details of the careers of geography graduates in Australia and New Zealand respectively;
- the advice and professionalism of Sage and Robert Rojek, our editor, for bringing this book to completion;
- our spouses for their unfailing support.

INTRODUCTION

Geography is the subject which holds the key to our future.
Michael Palin, broadcaster and author

1.1 WHAT THIS GUIDE IS TRYING TO DO

This Guide has been designed to help you study geography and related subjects at university. Whether you have already started or are about to start a geography degree, the Guide aims to let you make the most of your time in higher education. Whether geography (or a part of it such as human or physical geography) is the whole of your degree or a substantial part of it, this Guide shows you how to get more out of your time at university. We hope you will enjoy studying for your degree and become a better geographer and more employable. The information and advice here should be as relevant to part-time students as to full-time ones; and to those taking a full geography degree as to those following only a few geography courses or modules within a different degree scheme. So how is the Guide going to help you?

First, we want to explain what higher education, focused on geography, is trying to do and how it will help you develop into a resourceful, versatile and self-confident person (we discuss this further in Chapter 2). This chapter also discusses how to study and study more effectively in terms of what the research literature tells us about the different ways in which people learn things.

Second, the Guide tells you what qualities employers are looking for in their prospective staff, so you will know what to aim for during your three or four years at university (see Chapter 3 for details). The study of geography is unlikely to take up all of your time at university, nor should it. So in Chapter 6 the Guide suggests several ways in which you can take the initiative and use your spare time to enrich your period at university (and the rest of your life) and improve further your chances of getting the good job you want. Geography, we believe, can really let you get more out of life overall as well as being useful in career terms.

Third, the Guide de-mystifies the various elements of your degree and of the geography courses and modules which make up your degree. It explains why staff use devices like lectures and tutorials, examinations and essays; what they are using them for; and what you can do to get the most out of them. This is what we explore in Chapters 4 and 5. We believe that it will help you if educators tell you why they are teaching what they do, why they teach it that way, what they expect of you and what they expect you will gain from it.

Finally, the Guide provides you with a framework to help you measure your personal progress towards your goals. At various points during this Guide we shall talk about the value of assessing your progress and reflecting on how you are getting on. There is a grid in Appendix A which gives you a structure for this. In Appendix B there is a log where, under various headings, you can add to your personal record of achievements as your degree develops, year by year.

This Guide is necessarily short, which is no bad thing since you can read it quickly. It is *not* a complete geography degree course in one slim volume. It is an overview – something that is often missing – that shows you how all the components of a geography degree fit together. It suggests some steps you can take to make the most of your time at university studying geography.

1.2 THE GUIDE'S LIMITS

So, this Guide has been designed to operate within certain limits:

- it does not teach you geography as such; it is about how to study geography which is what the geography textbooks don't tell you;

- it deals mostly with how to study geography rather than with how to study in general, although many of the issues here are applicable to other subjects;
- it does not cover the 'lifestyle' issues of being a student (e.g. your social life, diet or sport);
- it provides general guidance on geography degrees and geography courses/modules, and obviously cannot deal with the unique features of individual geography departments.

1.3 MAKING THIS *YOUR* GUIDE

You are not 'just another student'. You are you; different from other students in terms of your current skills, your interests in life and personal values. These differences will affect how you interact with your geography degree. So we have written this Guide in a way that lets you 'customise' it. There are sections throughout the Guide where we invite you to pause and think about yourself, your academic progress and your personal development. Here is your first chance to do this.

ACTIVITY I

Try to get clear what your starting point is. You might like to jot down here your thoughts on five points.

1) Why did you come to university?
2) What do you hope to get out of your university degree?
3) Why did you choose geography rather than another subject?
4) What would you like to be doing five years after graduating?
5) As well as earning some money, is there anything else you would like to be doing or contributing to society, family or friends in five years' time?

1.4 HOW TO USE THE GUIDE

To get the most out of this Guide we suggest that you use it in two ways. First, it would be useful for you to *read through the whole Guide* fairly soon, to get an overview of the way we see geography degrees working. A geography degree has many elements which combine to form an integrated 'package' of higher education. This Guide shows you how the various elements of the degree combine and why staff use them. Second, you can use the Guide as *a reference work*, to be taken off your shelf whenever you need ideas on a specific topic (e.g. how to improve your essays).

If you are *just starting at university* or are reading this Guide before you go to university, then we would recommend that you focus particularly on Chapters 2 and 3, and Sections 4.1, 4.2, 4.4, 4.10, 5.1–5.4 and 6.1–6.6. Some other sections can safely be left for later, for example, Sections 4.3, 4.6, 5.6 and 6.7–6.10, since you are unlikely to be doing a dissertation or applying for jobs until much later in your degree course.

If you *have already progressed* some way through your degree, then you could skim quickly through Chapters 1 and 2, and Sections 3.1 and 3.2, whereas it is more important to read Chapters 4, 5 and 6.

For everyone, Chapter 7 is a very quick summary of the whole Guide.

Please remember that although we can give advice and guidance, there are no sure-fire routes to success. You need to read the advice and then adopt those aspects which suit you and your temperament. If a technique of study is working well for you, then carry on using it. However, if you are dissatisfied with your results, you can look through this Guide for alternative ways of learning which might give you better marks.

ACTIVITY 2

You may be assigned to a tutor or a member of staff early in your first year at university. That person's job will be to help you with university study in general and with the particular task of learning geography. A tutor is a useful person to get to know soon, and he/she will want to get to know you. So make sure you accept his/her invitation to meet and keep in touch. Your tutor could turn out to be really helpful.

1.5 FURTHER READING

At the end of the Guide the 'Further Reading' is designed to expand on what has been included here. You will find references to books and other materials on, for example, how to prepare a dissertation, how to produce a curriculum vitae or résumé and how to write essays. The books are all easily available in university libraries and some will also be found in major bookshops. Particularly useful is Pauline Kneale's *Study Skills for Geography Students: a Practical Guide* (London: Arnold, 1999).

In some ways learning geography is not all that different from learning many other subjects and so you may also want to look at some of the general 'study guides'. They are listed in full and alphabetically in the 'References' at the end of the Guide. Those by Barnes (1995), Marshall (1995), Northedge (1995), Becker (1986) and Rowntree (1998) are very useful. Tolmie (1998) provides a tantalising group of commentaries from students in a number of disciplines who got the very best degree results (first-class honours degrees); unfortunately, most of them could not fully explain why they were quite so successful at university!

So, what is geography?

Geography is about places:

- *what they are like now, were like in the past and why they changed;*
- *how the people and natural aspects of places affect each other;*
- *how and why economies, societies and cultures, and physical systems work in distinctive ways in different places;*
- *how local, regional, national and global systems interact.*

By the end of his/her degree a geography student will:

- *be fascinated by the diversity of places and understand how they work;*
- *appreciate as a citizen different traditions and environments;*
- *understand as a scientist a wide range of contemporary issues;*
- *appreciate the complexity of human and physical systems;*
- *be familiar with the social-science and natural-science approaches to studying the world;*
- *be critical of orthodox thinking and what we take for granted;*
- *be able to describe, analyse, research and understand places using different traditions of study;*
- *be able to communicate that understanding in different ways;*
- *have been prepared intellectually and in terms of abilities to study, understand and communicate a wide range of complex issues long after leaving university.*

. . . which is pretty impressive!

1.6 WHERE NEXT?

Of course, one of the most important points has little to do with geography specifically. You need to think about why you are at university at all, and the ways in which higher education changes people. That is what we shall consider in Chapter 2.

PITFALL I THE SHAPE OF THINGS

Which of these is the correct shape of Greenland?

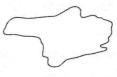
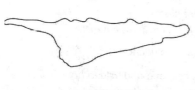

FIGURE I 'Shapes of Greenland'

None of them; they are all distortions. There can be no absolutely correct flat representation of a part of a spherical surface, such as the Earth's surface. Which distortion is best for your purposes?

No map is a perfect representation of reality. Similarly, words and statistics can never describe things perfectly – they always organise, highlight, select and distort the features of places.

A WORLD OF SHAPES.

FIGURE I *Transporter Bridge, Middlesbrough, England*

FIGURE 2 *Sunflowers and silo, Garden City, Kansas, USA*

2

WHAT GEOGRAPHY IN HIGHER EDUCATION IS ABOUT

There are some ideas so wrong that only a very intelligent person could believe them.
George Orwell (attributed to)

If you were to ask us 'why should I go to university and study for a degree in geography?' there would be six answers we could give; that is, six benefits you should gain from higher education. They are:

* learning about yourself and others;
* learning to learn;
* seeing things differently;

- applying knowledge, skills and the ability-to-learn to new topics;
- becoming independent;
- exploring geography.

In this chapter we shall describe these six benefits from higher education.

2.1 LEARNING ABOUT YOURSELF AND OTHERS

I have never let my schooling interfere with my education.
Mark Twain

Whatever subject you study, the most important thing you should gain from going to university is a deeper understanding of yourself. You will find out what your strengths are, the direction you want your life to take, and what personal and political values you are going to take into later life. You will gain this deeper knowledge of yourself not only because you will have had the space to get to know yourself better, but also because you will have developed as a person. Perhaps you are living away from home for the first time; meeting many new people; exposed to new subjects and ways of looking at the world; your established ideas will be challenged; you will have many opportunities to develop new interests. All in all, life at university, whether you are 18 or much older, will be an experience that is likely to change you. If you are a mature student, you may already have a clear idea about how you hope university will change you.

Of course, you can only make a successful choice of career if you really know yourself. Think about the following careers – teacher, accountant, research scientist, national park ranger, public relations consultant, project manager, European civil servant. Each of these jobs will appeal enormously to some students and be the last thing others would want to do. Many people have strong views on the kind of job they want – living in London or New York, in a town near home or overseas; individualist or team member; working for a commercial organisation or as part of a caring profession. To choose your career you really do need to know yourself and studying at university will help you do that.

Arguably, geography will help you to get to know yourself better than many other subjects because of the wide range of contemporary issues you will study in geography (human, physical and environmental). There is also a diversity of approaches to the subject (theoretical and applied;

natural science and social science) and many teaching m
in geography (see Chapter 4 for details). Geography
opportunities than most subjects to work in differen'
subjects, and that should help you learn more abou.
values. The more you know about yourself, the more selt-ᴄ
are likely to be in your abilities to learn and cope.

ACTIVITY 3

What would you say is your greatest strength as a student new to university?

How might you make best use of this strength?

If you suspect that you might have a weakness as a student that might hold you back at university, what is it and how might you minimise or work round it?

You will also get to know other people: staff such as lecturers, professors and tutors; other geography students in classes, tutorials and group projects; and students studying other subjects during your social life at university. This is important – understanding and getting on with all sorts of people is vital throughout your life.

2.2 LEARNING TO LEARN

*Education is what is left when you have forgotten all you have
ever learned.*
Anon

You may think you do not need to learn-how-to-learn because you already know – after all, you did well enough at school to get to university and you may have held down jobs after leaving school. To an extent you are right; you have already shown that you can learn things. However, at university you are going to be faced with a tougher task than at school or in the workplace. There will be more material to master, higher expecta-tions of you and more emphasis on you as an independent learner. *Under hese pressures you may have to re-think how you learn* – Section 2.12 could help you here.

You will have to find more of your own material to learn from; you will have to think more for yourself; and you will have to work out for yourself how to improve your performance. The ways of studying that saw you successfully through school or employment may not be sufficient for the harder job of learning at university. But think of the eventual prize – not only the ability to master geography to degree level, but also a set of learning skills that you can use on any task, however complex or obscure, during the rest of your life. So, learning-how-to-learn at the highest level will be one of the main benefits of a university education. This Guide aims to help you re-think how to learn geography.

So, if 'learning to learn' is a big part of doing a geography degree, how do you learn to learn? Clearly there is no one way in which all students will learn most effectively. We all have our own ways of working. If the way you organise your work is succeeding for you, then keep it going. However, if your results are disappointing, then you might like to consider other ways of working, such as the suggestions in this book (especially Section 2.12). When you get an essay back, review the mark and comments, noting what the lecturer liked and what was not appreciated so much. Try to focus on the weaker areas and see how they can be improved. That way you consolidate your strengths and work to improve your weaker areas.

The tricky part is that your various lecturers may react differently to your work. Some may put more emphasis on you showing that you know the relevant factual material, whereas others will be more impressed by your showing evidence of wide reading beyond the lecture notes. Some may be sticklers for correct spelling and grammar. Some staff will not tolerate the use of the word 'I' in essays; they say that essays should be written in an impersonal manner – 'it may be argued that . . ..' and not 'I believe that . . .'. Other staff will accept 'I'. To a limited extent – let's not exaggerate the variations among staff – you need to try to work out what individual staff like and expect from the students taking their courses, and then write accordingly. A good clue as to what staff place emphasis on is often found in their lecturing style.

Of course, you don't have to learn geography on your own. There is often merit in getting together with a few other students to work in a team, helping each other to understand your courses. If there are crises, you can support each other.

Traditionally, most undergraduates in geography have been full-time students; that is, they can attend university every day throughout the academic session. But what if you are not a full-time student, perhaps because you have family members to look after or you need to support yourself through university with paid employment? How then do you cope with fixed timetables, field courses and library systems designed originally for the full-time student?

Part of the solution may come from web-based courses and study packs which can be pursued at your own pace and at times which fit in with your other commitments. You may be able to get a fellow student to photocopy his/her lecture notes for you or to tape record the lectures. Your university library may make special arrangements for part-time students to borrow books for longer periods, but if they do not, you may need to reserve books ahead of time so that they will be available when you visit the library.

Field courses can be a problem – even one-day excursions let alone longer residential courses. Time off work and family care arrangements may be difficult and expensive to arrange. Your department should be open to the idea that you can develop your fieldwork skills in other ways, from your base at home and at your own pace. It would certainly be worth enquiring about this.

Some students with special needs (for example, those with restricted mobility) may find that the physical demands of some types of fieldwork are rather too taxing or unsafe. Again, the best advice is to talk through the proposed fieldwork, dissertation or practical work with the staff so that problems can be foreseen and worked around. The same advice applies to lectures and tutorials if you have impaired sight or hearing, for example. Staff can more easily help once they know your needs.

2.3 SEEING THINGS DIFFERENTLY: CHALLENGING ORTHODOXIES

What is this life if, full of care
We have no time to stand and stare?
W. H. Davies

One of the ways in which higher education can change you is that you learn how to see things in a new light. You can break out of conventional ways of thinking and create new perspectives. Two examples may help explain this.

Question

Two mothers and two daughters go fishing. Each catches one fish. How many fish did they catch?

Answer

'Four' is of course the obvious answer. But 'three' is also an acceptable answer. Can you explain why?

Explanation

If the group comprises a family group of a grandmother, mother and daughter, then the 'mother' is actually both a mother (to her own daughter) and a daughter (to her own mother); she fulfils both 'mother' and 'daughter' roles. So 'two mothers and two daughters' could be just three people.

Conclusion

The point about this question is that it shows how you can break out of conventional assumptions about what a situation means and how it should be tackled. These assumptions may be deliberately encouraged through the way that documents are written (so, 'two mothers and two daughters' suggests four different people). How questions are phrased can channel readers into a conventional way of thinking. For example, a problem with traffic congestion may be presented as simply a matter of building a new road and the only question is which route the road should take. Of course there are other ways of seeing the traffic congestion issue, using solutions such as public transport, flexible working hours and road pricing. Seeing things differently is quite liberating because it opens up new possibilities.

Question

Which way round do the hands of a clock go?

Answer

The obvious answer is that a clock's hands go round in the direction we call 'clockwise'. But this is not true if you are the clock. Can you explain why?

Explanation

The clock 'sees' its hands go round anti-clockwise because its viewpoint is the opposite to that of a person looking at the clock to tell the time. You can check this by taking off your watch (assuming it is a watch with hands) and holding it up in front of you so that you can see the back of the watch and someone facing you can tell the time. This gives you the clock's perspective on the world. Looked at this way, the clock 'sees' its hands going round anti-clockwise.

Conclusions

The first conclusion is that perspective matters. For example, the investor seeking profit by financing a hydro-electric dam will see the process of economic development differently from the farmer displaced by the rising waters of the dam's reservoir. Similarly, the hunter and the conservationist, and the traffic planner and the community group, may have diametrically opposite viewpoints. To understand the dynamics of the human world you have to appreciate all the different ways of seeing an issue. Some of these ways of seeing the world may initially seem strange, even to the extent of not occurring to you – like the clock's hands.

The second conclusion is that the question above initially seems ludicrous because the answer is so self-evident – what else would a clock's hands do but go round clockwise? It really does not look like a serious question. Wasn't that your first reaction to it? Hence anyone who asks such a silly question opens themselves to ridicule. So, answering questions can be difficult. But even more difficult is answering the questions no one has yet thought of asking. Challenging orthodoxies often starts by asking apparently odd questions – surprising ones or not what you expected.

So, two important lessons about higher education come from this. At university you will learn how to see things differently, from new perspectives, which can help you break out of conventional ways of thinking and framing issues. An analogy is with the child who asks the uncommonly sensible question. You will also learn that challenging assumptions and even asking new questions (or old questions in new ways) is possible and gives you a much fuller view of how the world works and how it could work in the future.

A more visual way of making the same point is the picture below. What is this thing?

FIGURE 2 *'White or black'*

What are you seeing here?
Is it a white urn against a black background?
Is it two facing profiles with a white area between them?

Well, of course, it is *both*. Personal interpretation is important in geography, as is the ability to see both the obvious structure and the less obvious one. It is also *neither* of these, since it is really just shapes on a flat surface, not three-dimensional objects like urns or faces. Yet geographers, like everyone else, use symbols to represent reality. Finally, it can be *whatever you want it to be* – geography as a subject has a creative element to it.

PITFALL 2 DIFFERENT DISTANCES

B   A   C

A is equidistant in miles from both B and C.

B   A   C

The A–C motorway brings C closer to A in terms of travel time than the slower roads to B.

B   A

C is inaccessible from A by train, unlike B.

B ||| A   C

The international border between A and B puts a wider cultural distance between these two cities.

There are many measures of the distances between places that can affect people's behaviour. So, pairs of places can be both 'near' and 'far' from each other.

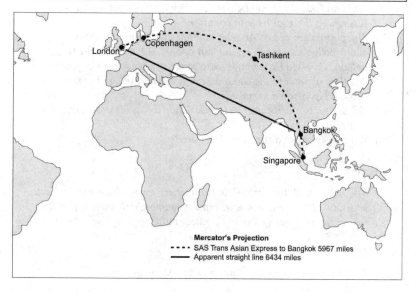

FIGURE 3 *'Routes to SE Asia' – the shorter route is the longer one*

A different sort of Geography quiz

A more directly geographical approach to this is through the following quiz. How many of these questions can you answer correctly? They all require you to 'see things differently'; they are *not* just about testing what you know (though knowledge is also important).

Questions

1) A square house has each of its outer walls facing south. Can you explain this?
2) Name two places on Earth where you could go skiing at the Equator.
3) In what sense is America always behind Japan?
4) Mauna Loa in Hawai'i (4170m) is arguably the tallest mountain in the world, not Mount Everest (8850m). What is the argument for this being true?
5) Why do they grow bananas in Iceland?
6) Henry was born in New York to parents who were both also born in New York. Neither Henry nor his parents were American. Can you explain this?
7) What do Oslo, Morecambe, Harare, Brighton, Chennai and St Petersburg (Russia) have in common?
8) The population of Los Angeles in 2000 was given in various authoritative sources as 3.7 million people, 9.8 million and 16.3 million. Can you explain this wide range of populations assuming that they all refer to the same date and urban area, and none of the authors has made a mistake?
9) Edinburgh is much closer to London than it was 50 years ago. How come?
10) What, apart from a currency, is a euro?

Answers

1) The house is at the North Pole from where everywhere else on Earth is 'south'. There are places where the seemingly unlikely is possible.
2) (a) Ecuador (near Quito, the capital, and Cotopaxi, the volcano – 5897m)
   (b) Kenya (Mount Kenya – 5199m)

   Avoid stereotyping – the equatorial is not always hot.

3) *The date in Asia is always one day ahead of the date in the Americas. There are many dimensions or criteria by which places can be compared, and not all will lead to the same judgement.*

4) *Mount Everest (Chomo-Lungma in Tibetan) at 8850m is the tallest mountain in the world if you measure the height of mountains from present sea level, which is the normal convention. If, however, you measure the height of Mauna Loa from where it starts its unbroken rise from the Pacific ocean bed to the volcanic crater at its summit, then as a physical structure it is 17km high (twice the height of Mount Everest), although only 4170m of this is above sea level.*
Conventions are useful for comparing things fairly but they can also be blinkering at times.

5) *Because the Icelandic tourist industry generates a plentiful supply of gullible tourists who are willing, just this once, to purchase an expensive Icelandic banana grown with geothermal heat. Abnormal combinations of circumstances can produce unusual outcomes.*

6) *They were all born before American Independence in 1776, so were British citizens. Think historically as well as geographically.*

7) *They have all changed their names. Oslo was Christiania, Morecambe was Poulton-le-Sands, Harare was Salisbury, Brighton was Brighthelmstone, Chennai was Madras and St Petersburg (Russia) was Leningrad (and before that Petrograd and St Petersburg). Cities, countries, ministries and firms all change their names. Tracking these changes can be difficult. Many name changes are a signal that some fundamental re-orientation or re-imaging of the entity is taking place, so it often signifies much more than just a change of preferred name. What we call things matters deeply.*

8) *It all depends on what you mean by 'Los Angeles'. Different sources use different boundaries, ranging from the City of LA (the small administrative area at the centre (the lowest figure)), through the County of LA (the middle figure) to the five-county LA conurbation (the highest figure). You need to define the terms you are using in academic debate; or, more simply, know what you are talking about.*

9) *The map distance between Edinburgh and London has not changed much, but they are closer in terms of travel time (faster roads and cars, trains and aircraft). This 'space–time convergence' applies to many pairs of cities but more to pairs of major cities than to pairs of minor ones. Distance can be measured in ways other than miles/kilometres and arguably travel time is at least as important to (potential) travellers as map distance.*

10) *A euro is also a small Australian wallaby.*

2.4 APPLYING KNOWLEDGE, SKILLS AND THE ABILITY-TO-LEARN TO NEW TOPICS

The reasonable man adapts himself to the world: the unreasonable one persists in trying to adapt the world to himself. Therefore all progress depends on the unreasonable man.
George Bernard Shaw

Learning geography is also about learning how to apply knowledge and skills to new situations. By the time you get to the end of your degree you will be able to show that, faced with subjects almost unknown to you when you entered university, you were able to master them to degree level. Having seen that you have done this once for your geography degree, employers will be confident that you can learn other new skills in whatever jobs they decide to give you. In short, that you are adaptable and capable of being creative and dealing with new situations.

This ability to apply knowledge and skills is increasingly important. If you stay in one job for a long time, you will find that the job will evolve; even to stay still, you will have to learn and apply new skills. Increasingly (as we shall explain in Chapter 3), you may find yourself moving (or being moved) between jobs. Again, it will be crucial for you to have the ability to learn new skills and new bodies of knowledge and to be able to apply to new jobs the knowledge and insights you gained from your degree and previous jobs. Mature students have some advantage here over school-leavers since they will already have experienced in their former jobs the need to learn new skills and adapt their existing skills to new situations.

2.5 BECOMING INDEPENDENT

It is impossible to enjoy idling thoroughly unless one has plenty of work to do
Jerome K. Jerome

Part of the experience of university is about becoming independent. What does that mean? It means:

- operating at a high level using your own intellectual resources in areas which are complex and not routine;

- being self-confident enough in your abilities and judgement to be effective and bold when working independently or with others, even on large projects;
- having the ability to weigh up evidence and situations;
- being critical of arguments and evidence;
- reaching fair judgements which you can justify;
- being able to monitor and improve your own performance.

Your geography degree will equip you in all these areas. You will benefit from having these skills and qualities not only for whatever jobs you go into, but also as a citizen – you will get so much more out of life.

2.6 EXPLORING GEOGRAPHY

Geography is a very rewarding subject. It covers a wide range of topics, giving you an insight into the environment, economy and society and how these three interact. It ranges across the continents, encompassing global issues and local events, and showing how they are connected. Geography studies why places are distinctive and how they have changed over time. All in all, geography is a superb university subject with plenty of scope for you to specialise later in your degree in specific areas which capture your imagination.

Only by studying a specific university-level subject – in this case, geography – can you acquire the deeper intellectual qualities and secure your personal development as we described earlier. Additionally, we would argue that geography is a better subject for you to study than many others because of its very wide range of teaching methods (which we shall describe in Chapter 4) and methods of assessment (see Chapter 5).

You might want to get a general overview of geography as a whole before you plunge into the detail of specific courses. Try dipping into some of the books which show the full range of geographers' interests and point out the important real-world issues in which geographers have become involved. These authors have also tried to demonstrate how geographers approach issues and think about them. Among the best such books are those by John Allen and Doreen Massey (1995), Peter Haggett (1990), Doreen Massey and John Allen (1984) and Alan Rogers, Heather Viles and Andrew Goudie (1992). The full references to these books are given in the References at the end of the Guide.

Learning geography is an excellent vehicle for you to achieve the transformation that higher education will create in you as a thinker, worker and citizen. But remember that there are many ways in which you can learn geography. You can learn:

- from your teachers (that is, from your lectures, tutorials and informal discussions);
- from and with fellow students;
- for yourself (reading books and articles, planning essays, revising for examinations);
- from your assessed work (preparing an essay, and reflecting on the marks and the feedback comments on your work);
- from reflection (thinking back on your geography, making links between modules and other subjects in your degree scheme, and considering the geographical importance of current affairs).

These are all equally important and useful ways in which to learn geography. Learning geography is not confined to formally timetabled periods of your life, such as lecture slots and field visits. Geography is all around you all the time, as you travel around town and from city to city.

ACTIVITY 4

Many people find that they can improve their understanding of geography by discussing a concept or issue with a fellow student. Working with a group of friends can generate new ideas and it complements working on your own. Studying does not have to be a lonely, friendless experience and it is probably more effective and fun if it is not. Indeed, working with others is itself a key skill.

2.7 GEOGRAPHY AT SCHOOL AND UNIVERSITY

As you prepare for the move from school to university, the question arises of how studying geography at university compares with the subject at school. Is it much the same but a bit harder? The answer is: 'No, it's really quite different.'

The distinction is partly between *being taught* at school and *learning* at university. At school many of the critical inputs to your education came from your geography teacher and what you read in one or two geography textbooks. At university the lecturer is not a teacher in the school sense of the word. Rather he/she is a guide to the subject, describing its layout but not giving you all the detail you will need. That detail has to come from the reading you do after each lecture and to help you prepare for essays and examinations. The phrase 'reading for a degree' describes precisely what you should be doing – reading widely and not just one or two textbooks. Your lecture notes (however thorough they are) are no substitute for that reading.

The difference between school and university goes deeper than the shifting balance between being taught and actively reading geography. The texts you will read at university are different in type from those at school as well as there being many more of them. They are less dogmatic (in the sense of telling you the right answer and what to think and how to see issues). They are more concerned with arguments, disputes and different ways of viewing geography. This is particularly true when you read articles in journals. These often advance new ideas and criticise other authors. Some schoolbooks are politically very neutral, and deliberately so; some articles are clearly not. Your task is not to memorise the right answers and facts (though the latter can come in useful), but to be able to understand, describe and evaluate the arguments and controversies and the conclusions the authors draw from their evidence and theories. In geography at university you will be presented with different authorities' views on how rivers erode, for example, how and why cities grow, and where economic development takes place and with what consequences. You will come to appreciate that textbooks and articles are the 'creations' of their authors and not just absolute statements of the truth. Geography at university is about you learning to handle uncertainty and debate, contributing your own arguments and finding your own views on geographical questions – quite liberating, really. Section 2.3 goes into more detail on what it means to see things differently.

PITFALL 3 SPACES AND PLACES

A grid square on a map is a *space* – objective, real, universal and emotionless.

A *place* is somewhere that has meaning for individual people or groups because of its associations, such as Mecca, Gettysburg, Bannockburn, the specific school you went to or the venue of your happiest holiday.

Places have power; spaces just exist.

2.8 WHAT SORT OF SUBJECT IS GEOGRAPHY?

Of course, you already have a good idea of what geography is, based on your work at school. At university the courses will build on this foundation. Most geographers see the subject as having a number of recurring themes which distinguish it from other disciplines and help unite the human and physical aspects of the subject. Seven themes can be identified. These were set out in the UK Quality Assurance Agency's *Geography* Benchmark Statement (2000) which was written by a group of British geographers to describe the subject's common view of itself.

Scale

All processes unfold at a variety of scales, from the local to the global, and what happens at one scale often affects events at the others. So, global trends (for example, climate change and environmentalism) affect local areas, and local events can have global consequences (e.g. Chernobyl). Geographers study these scale effects in both human and physical geography.

Space and places

Space is just distance (measured in kilometres, or from a market or city centre). Places are bits of space that have been given personal or social meaning by people – Hiroshima is not just another Japanese city; Lourdes is not just another small French town. My first school and your home street have meanings to you and me that other schools and streets do not. They are our places as well as spaces on the map. Human geographers study how places acquired their images and meanings, and how these can be altered if necessary, and viewed differently by others – is it Londonderry or Derry?

Systems

Everything is related to everything else but some things are particularly strongly connected. Events have effects – drought and farming; road building and economies; branch plants and headquarters; discrimination and urban communities. It needs geographers' analytical skills to distinguish between those things that merely co-exist and those that affect each other.

Spatial variation

Although some processes and systems operate widely, they do not operate in the same ways or with the same consequences everywhere. So, industrialisation was not the same in France, Japan and Mexico because in each place a unique set of factors modified the general process. Neither will global warming have the same predictable effects all over the world. Events are contingent on their location, and processes vary spatially. So, explaining the spatial variations we see all around us is an important part of geographers' work.

Environments and landscapes

These two elements typify the human/physical linkages in geography. Landscapes matter because of how people react to lakes and mountains, not just because they exist as real objects. Environments matter because of how people change them and the human and other consequences of such changes. Some geographers study how people interact with their environments.

Change

There is a strong historical dimension in geography. Everything (every valley, industrial district, tourist resort, habitat) is the result of processes

LOOK OUT FOR IMAGES, ICONS AND STEREOTYPES OF PLACES.

FIGURE 3 *Skyscrapers and reflections in Denver, Colorado, USA*

FIGURE 4 *A Swiss chalet*

(erosion or economic development, for example) which have evolved over time, sometimes predictably, sometimes in unexpected ways. How geographical features have changed, and have been changed, over time is a feature of many geographical studies.

Differences

Difference refers to the inequalities among groups of people, how these arise, their consequences and how they might be reduced. Some geographers will work hard to reduce such inequalities.

These themes have traditionally appeared throughout geography and all its branches. Yet much else in geography is dynamic and debated. The recent history of geography (and of many other subjects too, of course) is rich in controversies over how the subject should progress. To what extent is geography an art or a science? Are there laws governing geographical processes or is everything so unique that models of change are misleading? What kinds of evidence are valid – quantitative or qualitative?

– or do we need only theories and no empirical research at all? Is the geography of an area the product of deeply embedded structures or of a myriad of individuals' actions? How should we conceptualise the linkages between people, groups, firms and governments, and between micro-scale and macro-scale forces in the physical and human realms? You will learn about these debates during your time at university. Rigid ways of thinking and orthodox types of research have never been under so much attack. Overall it is a fascinating time to be a geographer.

PITFALL 4 EMOTION

'Emotion has no place in geography'. Discuss

Many elements of geography can and should be determined without the use of emotion – the speed of the river's flow, the rate of coastal erosion, the number of farmers offering bed and breakfast accommodation, the literacy rate. Other aspects of geography – the tourists' view of landscape beauty, the morality of riches and poverty, the world-view of the immigrant woman in the slum – necessarily involve assessing emotions and probably require us as researchers to make judgements on the rightness of the situation. In these cases we need to be open about our position and background as researchers. Dangerous emotions are those slipped in and which sway the argument subtly – 'mass tourism' (sounds unpleasant); 'environmental destruction' (or is it just environmental change?); 'economic growth' (for everyone? in quality of life and incomes?); my use of the word 'slum' (above) when all I meant was 'poor quality housing'.

No academic subject is totally different from all others, but there are some features that are more fully developed in our subject than in many others. Geographers tend to be practical people (which may be why they get jobs easily), who are able to apply their ideas and skills to a wide range of aspects of the real world. Many geographers have an informed concern about aspects of that real world today – poverty or conservation, for example. Geography teaches you useful research skills (which is good for your career) and also about the big issues and debates without which practical research lacks purpose and direction. Geographers now use several different ways of reaching an understanding of the world and they tend to be good at communicating that understanding in writing and by other means. The subject is certainly a broader one than many others – you will not be bored by it – because geography is all around us.

Geography degrees usually begin with broad survey courses in Year 1 and end with specialist research-focused courses in the final year. Paralleling this progression in courses will be a steady growth in the skills you will learn until, by the end of your geography degree, you will have mastered a wide range of skills to an intermediate level and a few to high levels. The advanced skills and final-year courses may be concentrated in a particular part of geography (for example, human geography, less developed countries, the environment) as you shape your degree to play to your strengths and interests and/or prepare yourself for your career area.

2.9 ASSESSING YOUR PROGRESS

What do you want to get out of your geography degree and how can you be sure whether you are achieving your goals? It is useful – though very difficult – to try to work out the position you hope to have reached at the end of your geography degree. Then you can check how much progress towards that ideal state you have made so far and you can plan how to achieve your final position by the time you graduate.

The self-assessment sheet in Appendix A suggests a list of things you might have achieved so far. Turn to Appendix A now and try filling it in. Put a tick against those things or experiences that you have occasionally done or had, and put two ticks against those you are already pretty good at. There will probably not be many ticks if you are a school-leaver and you are doing this self-assessment right at the start of your university degree, but each year you should be able to add some more. It is worth repeating this exercise each year. By the time you leave university with your geography degree, most items will have been covered. *This is the path along which your university is planning to take you.*

Another useful activity to look at now is Appendix B – your Personal Record. Since a lot is going to happen to you during your time at university, it will help to get into the habit of recording *briefly* your achievements and progress at least twice a year. This will be very useful as a source of information that you can use to construct your curriculum vitae or résumé, which you will need when applying for jobs in later years.

You may think that, at the start of your university degree, your personal record would be rather empty. Not so. At school you will have acquired some skills in writing essays and you may have some experience of using computers. Holiday jobs may have taught you a thing or two. Your family, social life and school work will have given you some track record in working with other people to achieve a common goal. Of course, if you are coming to university rather later in life than the average school-leaver, you will have a head-start in personal skills from your former working life. So we suggest that you start jotting down your achievements so far. Your university may have a system of progress files like these that it asks you to use. If you compiled a Record of Achievement, Progress File or something similar at school, don't let that lapse; the basic idea behind such a personal record is still sound.

2.10 HOW HARD SHOULD YOU WORK AT UNIVERSITY?

This is a tricky question. The answer depends on how hard you need to work to get all your tasks done (some people work faster than others, though not necessarily any better). It depends also on your priorities. How much time for sport, hobbies and socialising do you want? If you have a paid job while at university or are looking after a family, those demands on your time also have to be taken into account.

As a guideline for a full-time student, try to work a 40-hour week in term time – the average full-time employee works about this long. You will probably find that about 15 hours per week will be timetabled – you have to be in a lecture theatre, laboratory or seminar room for a set teaching activity. The other 25 hours per week are for your independent reading, preparing coursework (e.g. essays) and working on projects. If some of your tasks involve group work with other students, then you will have to mesh part of your working timetable with theirs. Otherwise, when you do these other 25 hours' work is entirely up to you. You will set your own work pattern at university.

The vacations are for holidays and paid work but you can use them for project work and to catch up on reading. This use of the vacations can be very helpful for those with family commitments or special needs, allowing a more even pace of work over an extended period.

Arguably good time management is as important as the total number of hours you work – working smarter rather than longer. Except in an emergency, very long hours are often less productive than short bursts of work with relaxation time in between. As a guide, you should be able to write an essay in about a week. Day 1 is for thinking about the topic and finding the items to read in the Library or on the Internet. Days 2 and 3 are for reading the material and taking notes. On Day 4 you can plan the essay and on Days 5 and 6 draft it. Leave Day 7 to review and polish the text. So, you can work out how many essays and other pieces of coursework you have to complete by the end of the term/semester and pencil each one into its own week according to the deadlines set by your tutors. The thing to avoid at all costs is having to write six essays in two weeks!

Your dissertation needs the same sort of planning, working backwards from its submission date. The last week is for final printing and emergencies. Then schedule in the writing of the various chapters and the fieldwork and, first of all, the background reading. The earlier you do your sketching of the methods and chapters, the sooner your supervisor can advise whether you are on the right lines – writing a 12,000-word dissertation in a fortnight is tough! Unlike an essay, a dissertation is prone to unforeseeable problems which could slow you up or divert you. Dissertation planning needs slack in it to let you cope with contingencies – you have to expect the unexpected. So don't leave things to the last minute, if you can help it.

PITFALL 5 WORDS AND THINGS

Some words describe what we can safely call 'things' – rivers, cars and houses, for example. These are 'things' because they really exist and can be described with reference to external rules, standards or widely accepted conventions as to what a car or house is. There may be some debate at the margin – is a university hall of residence a house or does it belong in a different class of communal accommodation, alongside prisons, hospitals and hotels? However, other words do not describe such 'concrete' things – for example, 'culture', 'development' (sustainable or otherwise), 'progress' and 'modernisation' – because these are concepts that need to be defined. For the academic, what we (and others who use these words) mean by them is the interesting and important question. Many definitions of these terms are possible and in fact are used.

Words are words, given meanings by their authors and interpreted with these or other meanings by readers. In literature, a text with multiple meanings (your interpretation of it, mine and our colleagues' interpretation) may be exciting – the text means what each of us wants it to mean. What the author meant may be unnecessary to know and indeed ultimately unknowable with certainty. In geography, however, we often need to be clear that we are all using the words with the same meanings if we are to have a debate or discussion which leads to secure conclusions. So, what exactly does the author of the textbook you are reading mean by 'the Green Revolution'? Does he/she tell you?

2.11 THREE BIG DON'TS

This book tries not to preach or be prescriptive. But there are three things the undergraduate geographer really should try to avoid because they nearly always have a negative effect on people's performance, sometimes seriously so.

Don't be a perfectionist

If you set yourself unattainably high standards, this can lead to constant disappointment and a perpetual sense of failure. Of course, do your best, learn from your tutors' feedback, and try to improve. But then stop.

Don't over-emphasise the importance of intelligence

Some people see themselves as inherently intelligent (and so don't need to work hard) or as basically limited in intelligence (and so incapable of rising above some 'natural' level of achievement). Both groups are fatalists – their perception of how intelligent they are completely rules their performance. This denies the possibility that hard work and reflection can play a part in raising people's performance, whether they start from a high or low base level of intelligence. It is better to be optimistic that improvement is possible, than to be a fatalist.

Don't rely on lecture notes

Read and think for yourself; that really will impress your tutors.

2.12 HOW DO YOU LEARN AND HOW CAN YOU IMPROVE THIS?

You have probably been studying and learning for many years, but have you ever stopped to consider how you do this? Everyone takes for granted that they understand what learning is and how they set about it. When things are going well in your studies, *how* you are learning does not seem to matter very much. However, doing an undergraduate degree makes many demands on students; this may lead you to question how you study and learn, particularly when you come up against difficulties or face possible failure. The more successful you have been in your studies before university, the greater the shock if you find that your marks are not as high as expected. It is likely that the activities you associate with learning – reading, writing, working on problems, thinking, planning, practical work – may be rather different from how you experienced them in the past. For some students this change of gear happens in their first year. For others the transition to the second year is the point at which they begin to feel that they are being stretched – that the quantity and nature of the work they are being asked to do feels very different or more demanding. The transition to second year may be particularly problematic when, as in many universities, it is the work from the second year onwards that counts towards the class of your final degree.

At this point it is worth thinking about your approach to learning.

ACTIVITY 5

How do you typically learn?

- Do you need to see the 'big picture' before you can understand a new area of study or do you find it better to build up an area of knowledge bit by bit?

- Do you need to have practical examples before a particular concept makes sense, or are you distracted by illustrations?

- Do you like to explore a subject in depth before you can write about it, or do you quickly find that too much reading results in your feeling swamped and so less able to write?

There has been a lot of research conducted into how people learn. Some of this research suggests that a person's psychological make-up determines much about how he/she addresses learning. Others would argue that the situation in which you find yourself has a more significant effect on your approach. It may help you to understand a little more about how you operate if we now look at some of this research.

Learning styles

Honey and Mumford (1982) suggested that there are four learning styles:

- activists;
- reflectors;
- pragmatists; and
- theorists.

Activists are people who need to **do** in order to learn. They are bored and frustrated by overarching theories and need to get on with things. They tend to be energetic in their approach to learning and tend to get things done on time. This approach can count against them, of course, because they may have difficulty seeing the structure of the subject and may mistake activity for learning.

Reflectors need time to think. They like to mull over ideas and put off coming to conclusions until they really have to. As students they are likely to want to do a great deal of reading before committing themselves to writing or to contributing to a discussion. Being reflective is a valuable characteristic in learning at university. However, reflectors can often find it very difficult to get tasks finished. They may be paralysed by their search for the better, if not perfect, answer. They are likely to find planning problematic. They may also find it difficult to reach conclusions – essential in essay writing or experimental work!

Pragmatists are interested in how things work in practice. They have little interest in abstract theory, particularly where they cannot see its relevance to the real world. Pragmatists are likely to be quite strategic in organising their studies: rather than trying to do ground-breaking work, as reflectors or theorists may do, they make judgements about what is important and what can be left.

Theorists need, as the name suggests, to be able to see patterns or to organise what they encounter into theories. They tend to be unhappy doing apparently disjointed things and need to see the 'big picture'. People with this learning style will want to know what the structure of a course is before they embark on it. Not only that, but they will want to know **why** the course is organised in this way. Theorists will need to have this information from course handbooks or websites, or to have good and regular interactions with tutors to give them a sense of how a course is developing and where its individual elements fit into the big picture.

In practice, of course, no one adopts just one of these styles, and most people find they are using different combinations of styles according to the kind of task they are being asked to do. However, these descriptions can be quite useful as a framework to help you think about how you typically approach your learning. Each of these styles has its strengths and weaknesses. Judging which one to use on any occasion is the key skill.

ACTIVITY 6

Think about a learning task you have recently undertaken (e.g. the tasks of a summer job, learning to drive, learning geography for your A-level examinations). Can you recognise yourself in any of the descriptions above? If they do not seem to describe how you addressed it, how would **you** describe your approach? What do you think were the strengths of that approach in relation to the specific task you were undertaking? What were its weaknesses? Do you think you could develop a better way of doing that kind of task, should you meet it again in the future?

Self-perception

Another area that researchers into learning are interested in is the way that people view themselves, on the basis that our self-perception affects the way we act. In simple terms, if we start off thinking that we are able easily to do a range of things, the chances are that we will be able to do them. On the other hand, if we think we are hopeless at particular things, then the chances are that we will be. For example, you may well think, as many do, that you are hopeless at mathematics – the very word sends a shiver down your spine. There could well be some justification for this self-perception – possibly you have scraped through every examination in mathematics you have ever taken. However, some researchers would say that the very fact you have these negative perceptions about your mathematical ability is itself influential in predisposing you to poor performance in mathematics. And conversely, people are more likely to succeed in tasks if they approach them believing that their success elsewhere should lead to further success here.

Dweck (2000) has done a lot of work on what she terms 'self-theories'. According to Dweck, people subscribe either to a theory of intelligence that this is a fixed entity – this is called the entity theory – or to an incremental theory, which takes intelligence as something that can change and develop over time. Our behaviour and thoughts will betray whether we tend more towards one or the other of these positions. In the example above, the person who thinks he/she is useless at mathematics and that nothing can be done about it, is showing that he/she tends towards an entity theory of intelligence, at least in relation to maths. This is a very fixed position to be in. By firmly believing that ability in mathematics (or indeed any other area of study) is fixed and that you either have it or you don't, a person is limiting any potential he/she may have to develop. Those who subscribe to the incremental theory of intelligence, on the other hand, tend to remain optimistic about the possibilities of development and do not discount the likelihood that through further effort they may improve in particular areas they have previously found difficult.

It is easy to see how the entity view of intelligence may be of great benefit if you believe (backed up by your experience and the views of others) that you are brilliant in particular areas. Progress here is assured and should be rapid. However, it can also be a problematic position to take, in two ways. First, if you have always been good at something (say, geography at school) and then encounter difficulties (with, say, geography at university), you may conclude that you had previously been wrong about your intelligence in that area and feel like a terrible failure, possibly even wanting to give up. Secondly, you may tend to avoid things that you are convinced you will be no good at, and thus cut off possibilities for development in new areas. University degrees and university life in general offer many opportunities to study or do new things. Having a fixed idea about whether you are likely to be good at them may inhibit you from making the most of the possibilities that are open to you. You end up always playing to your strengths (a good tactic, of course) but never tackling your weaker areas because you firmly believe that you cannot improve them. The incremental theorist is convinced that thought and effort can lead to some useful progress even in demanding tasks. We, this book's authors, side with the incremental theorists, of course.

ACTIVITY 7

What do you believe about your abilities? What do you think you are good at? Why do you think you might have difficulties with other areas? What do you notice about the way other people talk about their ability or lack of ability in particular areas?

Knowing the game

Other theorists have focused more on the **context** of learning than on individual temperament or tendencies. In a classic text from 1974, Miller and Parlett talk about students being 'cue-blind', 'cue-conscious' or 'cue-seeking' in relation to success in their degrees. 'Cue-seeking' students are those who have worked out that getting through a degree successfully is a kind of game and that they need to spot the 'cues' about how to perform. These students will make sure they are well aware of regulations, of exactly what their tutors are looking for, and of what constitutes good work. These students will check out the essay banks in their Students' Union advice centre, make sure they get to speak to their tutors, and find out from last year's students what to watch out for. This is the academic version of being 'streetwise'. 'Cue-conscious' people will act in a similar way, although not perhaps to the same extent as the cue-seekers, picking up the cues but less actively seeking them out. Those who are 'cue-blind' will not have realised that being a student is a game, with implicit rules and codes. They are unlikely to pick up what is important and what less so. For them the danger is that they may put effort into activities that are unlikely to result in success (e.g. reading far too much or the wrong kind of material for an essay) or fail to recognise what is being sought in a particular task.

It may sound rather cynical to say that studying for a degree is a 'game', but it is important to recognise that, to some extent like all human activity, it does resemble a game. As in any game, a degree programme has a structure and a timetable, a beginning and an end; there are rules to follow and penalties for not doing so; there are expectations of the staff and student players. Taking some time to work out just how studying for a degree works will pay enormous dividends in the long run and help you to focus on what is likely to bring about your success when the final whistle blows – your graduation.

Learning from failure

It may seem strange to raise the issue of failure in a book about how to get the most from your geography degree. When you come to university you are likely to be thinking of yourself as something of a success: you have passed examinations, gained good grades and are ready to get started on a new challenge. You will be doing work which you hope and expect will get good grades. However, the path to a degree is sometimes not as straightforward as you might hope. The challenges may feel overwhelming at times, particularly when you may be coping with the new experience of living away from home and having to fend for yourself, or possibly trying to juggle home, family and paid work at the same time as your studies. Academic work will itself present previously un-met challenges – you will be required to be more independent in your work and to take on intellectual demands that may be quite taxing. You are quite likely to experience some form of failure during your degree programme because 'failure' does not just mean getting very low marks. You may *feel* like a failure if you get a mark you are not happy with (because it falls short of your expectations), or you do not get work in on time or do not contribute to groupwork in the way you know you should. Dealing with this failure (or fear of failure) is one of the things you need to learn. Some able students drop out of university, despite technically passing all their examinations, because they *feel* they have failed to meet *their* expectations of their performance. Perfectionism is a dangerous standard for students to set themselves – it almost guarantees a sense of failure.

If you can, look at failure in a positive way – hard, we know, but worth a try! If you consistently do well at your work, you are unlikely to learn very much more about yourself and how you learn. Feedback from assessed work is likely to be briefer for highly successful work than it is for work which has scored very low in the marking system. By doing less well on occasions you may find that you get indicators from tutors about which areas of your work need attention and what you may do about them. This is useful for the future. When you leave university after your degree you will certainly have to deal with the possibility of actual or relative failure in many areas of your life. By adopting a positive attitude towards failures and setbacks – I must learn how to stop this happening again – you will be gaining a very valuable skill. Try the following checklist to help you 'learn from failure'.

- What *exactly* went wrong here? Try to avoid thinking that *you are* a failure and focus on what specific thing you did (or did not do) on this occasion to bring the failure about.
- Look at any circumstances which might have contributed to the failure: which were within your control and which were outside it?
- What plans might you make to avoid this sort of failure in the future? These might include planning your time a little better, being more careful about getting hold of the materials you need, booking some time with a tutor for more detailed feedback, or seeing someone in your study-support centre well before the deadline for the next piece of work.
- Try talking to someone – a friend, a counsellor – about how it felt to fail. This will help reassure you that it is not *you* who is a failure. All that happened is that on this occasion you did something less well than you hoped and next time you could do it better.
- Remember that many great discoveries and innovations have come about because things went wrong and the people involved looked at what had happened and learned from it.

2.13 COPING WITH STRESS

Occasionally the pace of life at university may get oppressive – some illness, a flurry of essay deadlines and a relationship going sour all at the same time, for example. Here are some proven ways of coping when the going gets tougher.

- Deal with things one at a time. Focus on one, do what you can with that one and leave the others in the background. Coping simul-taneously with several problems is really difficult.
- Try to gain time by getting extensions for some of your coursework – even a few extra days will help.
- Do jobs in less time. You *can* write an essay in two days rather than over the week you would normally take. There will have to be less reading and it will have to be written without time for revising the text, so its quality may suffer somewhat. But if the choice is between a poorer

essay and one that gets a mark of zero because it is handed in late, then go for the poorer essay! And start your next essay earlier.

* Accept what you really cannot change or control. If it ain't meant to be, then you can't make it happen.
* Laugh it off or shrug it off, and move on. Learning from the past is good; dwelling on the past can be destructive and time wasting.
* Accept that obstructive people may have a different way of seeing things from you. What is it? Can you work with it or round it?
* Focus on what is really most important to you rather than on everything; few of us achieve all we want.
* Balance your life – all study is as bad as all fun.
* Make an early start on tasks. Leaving things to the last minute is stressful and, although some stress can boost performance, it leaves you vulnerable to problems, with little time to sort them out.

However, to appreciate fully the approach this Guide takes to learning geography, you need to be aware of the major changes which are occurring in the economy and in the kinds of graduate that employers are looking for. The next chapter describes and explains what is happening and how it will affect you after you have graduated.

3 GEOGRAPHY, GEOGRAPHERS AND YOUR FUTURE CAREER

In this Chapter we shall:

- discuss the traditional and more modern views on the links between university degree courses and graduates' careers;
- show why firms and governments both want more skilled graduates;
- describe the qualities employers are looking for in their future employees;
- demonstrate the particular usefulness of a geography degree today;
- discuss how to get the most out of a gap year.

3.1 WHY STUDY A SUBJECT?

One of the ideas behind this Guide is that you can use your time at university to make yourself more employable by gaining skills and experiences that will be recognised by potential employers as likely to make you a more effective member of the workforce. If we were writing a guide for those studying English, history, politics or many other subjects, the idea of university as a preparation for the world of work would be promoted just as strongly these days.

However, as late as the 1970s this notion that a degree subject should be mainly vocational would have seemed an odd one among students and would have been unacceptable to academics. For students, university was about studying a subject because one was interested in it, and also having a really good time during the latter stage of growing up into adulthood. For academics, the purpose of studying their subject was intrinsic and 'liberal'; you studied a subject for its own sake. If the subject turned out to be useful in later life, that was a bonus – it was not the university's purpose to train you for employment. University was about a general intellectual maturing, about transmitting liberal cultural values to a new generation, and about creating a group of critical, highly trained leaders for the future. That is all still true – geography is inherently interesting and you will benefit as a citizen from having studied it at university. But that is no longer the whole story.

3.2 RECENT CHANGES

So much has changed. In the early 1970s about 12 per cent of school-leavers in the UK went on to higher education and older students were rare. Today the figure in the UK and New Zealand is between 34 and 40 per cent of school-leavers and in the USA over 50 per cent. Additionally, mature students are much more common than previously. This trend from higher education for an elite to a mass system of education is not unique to these countries. It is found to a greater or lesser extent throughout the developed world and represents a sharp acceleration in the steady upward trend since the 1930s.

There are no longer enough jobs among the country's leaders and the traditional professions to absorb all these graduates. Graduates are now taking formerly non-graduate jobs. That might sound retrograde; it is not, however, because of other developments in the economy. Many public and private organisations are 'de-layering'. Having shed their production-line staff through automation, they are now thinning out the many traditional layers of management and bureaucracy. The resulting 'flatter' organisations – some with only five grades of staff between the most junior employee and the managing director – now need their quite junior staff (perhaps only recently recruited) to have some of the resourcefulness, intelligence and originality of thought which once were the hallmarks of a few high-flyers. High-level skills, such as the ability to shoulder responsibility and to solve problems, are now needed throughout most organisations. There is a real graduate's job to be done in traditionally non-graduate posts.

A further dimension is the role of government. Across the world, governments are looking at how they can boost the international competitiveness of their firms. The idea of using the national education system to make the country's firms more effective – because they have better trained staff – is obviously an appealing one. Why, it might be argued, should the taxpayer finance people to spend three or four years at university studying a subject of no use whatsoever? Why should the government not see their invest-ment in higher education as an enormous training programme for future workers? This utilitarian notion of what universities are for fits neatly with the concern among firms (described earlier) to recruit new staff ready to shoulder considerable responsibility soon after appointment, rather than after 20 years of unblemished service. Of course, if you are paying your own university fees, a degree course which promises to be vocationally useful may be a particularly attractive idea.

3.3 YOUR ROLE AND WHAT EMPLOYERS WANT

So, where do you fit into this picture? Are you just a high-grade operative helping your country's international competitiveness and bolstering the profits of 'slim and trim' firms? Partly, yes, but there is a lot in these changes for you. You may get more interesting and responsible jobs much earlier in your career. You may wish to move between jobs more frequently to get a higher salary, or you may be moved by your employer as they restructure the firm, closing down some operations and starting up others. Careers may be about to become more fluid, with people having to, or wanting to, change job more often. You may need to re-train mid-career rather than the traditional idea of a life of employment with a single employer after graduation at the age of 21.

This may actually be what will happen, though the statistical evidence is not yet clear. It is undoubtedly true that governments and employers have been pressurising universities to alter their courses away from the purely subject-centred (you go to university to be taught astrophysics or Anglo-Saxon) to being more skills-centred. Initially, this took the form of a major concentration on 'skills training' of a fairly low-level character, such as the ability to handle numbers, to write fluently, to give public presentations and to manage one's time. These were the immediate deficiencies which employers claimed to find in their newly graduated recruits, and the universities' job was to put this right. A similar set of goals existed within the school system.

Gradually there has developed a more considered view of the qualities that employers and employees think will be needed in the future. The aim now is for a more complex set of abilities and experiences among graduates. You will now need:

- to be able to solve problems;
- to apply ideas and skills to new areas;
- to be able to work with others (to argue, negotiate, co-operate, compromise and win);
- to be self-reliant;
- to be able and willing to learn throughout your working life;
- to be able to master a wide range of complex topics;
- to be able to cope with uncertainty and change;
- to be self-confident;
- to be self-reflective (able to improve yourself without teacher/boss always having to be present to direct you).

Perhaps now you can appreciate why, at the start of Chapter 2, we said that among the benefits of higher education were:

- learning about yourself and others;
- learning to learn;
- applying knowledge, skills and the ability-to-learn to new topics;
- becoming independent.

It is these higher-level and less easily definable qualities which will be the most important for your personal career progress and for your employers.

If you think about this new vocational agenda for higher education from the viewpoint of the universities, you can quickly appreciate their problem. It is easy enough to see how you can build, say, statistics training or fieldwork into the degree scheme so that geography graduates will be numerate and practical (particularly when class sizes are small). But how do you make large intakes of students self-confident or self-reliant or good problem-solvers? These qualities and experiences take a great deal more ingenuity on the universities' part to incorporate into degree schemes. The universities' problem is made all the more acute by the fact that staff:student ratios are rising – there are fewer staff employed to teach more students – and staff are under increasing pressure to spend more time doing research or generating income. The result is that more emphasis is being placed on you as an alert learner actively quarrying your geography degree course for all that it can give you.

ACTIVITY 8

Buy or borrow a copy of one of the quality newspapers – one with job advertisements for graduates. Look through the jobs pages and draw up a list of the jobs which mention the skills you can expect to develop as a geographer. What sort of jobs are these?

3.4 GEOGRAPHERS' CAREERS

Employers rarely need narrow specialists. They need bright, committed, inquisitive and determined individuals with the ability to draw on what is happening elsewhere in the world. The best geographers are stars in this respect!
Professor David Rhind, Chief Executive (1991–98), Ordnance Survey (the UK's national mapping agency)

Being a specialist is one thing, getting a job is another.
Stephen Leacock

Why do geographers have one of the higher rates of graduate employment? What do we know about the careers that geography graduates have actually followed in the recent past? Their experiences have varied over time (for example, recruitment into the financial services sector has waxed and waned) and they have differed from country to country, but some general points are clear.

Careers start soon after graduation but they then evolve: sometimes students switch jobs after a few years, perhaps following the completion of a postgraduate degree or a re-appraisal of longer-term goals. In 1998 the Royal Geographical Society (with the Institute of British Geographers) surveyed the careers of a sample of British geographers five years after they had first graduated (see their website at http://www.rgs.org/ed/). The six major career areas for geographers were:

- administration and management;
- teaching and lecturing;
- the financial sector;
- marketing;
- research;
- industry and manufacturing.

It might be more helpful to give more detail on specific jobs rather than broad areas of employment. Here are some of the specific jobs and further training obtained soon after graduation by the graduates from one British geography department in recent years.

Further study (apart from teacher training and higher/research degrees)

Masters courses in Town Planning, Transport Economics, Environmental Policy and Regulation, Historical Research. Certificate in Outdoor Education. Diploma in Radio. Law Common Professional Examination.

Administration and public service

Administrative officers with Serious Fraud Office, Benefits Agency, Courts Administration, British Telecom, British Gas, Legal Aid Board, National Farmers' Union, Countryside Agency, BUPA, Ministry of Agriculture. Legal administrator. Various Civil Service posts. Town and Country Planning. Transport Planning.

Marketing and sales

Trainee managerships with major supermarkets, high-street stores, food companies, car companies and department stores.

Service occupations

Teaching (primary, secondary and further education). Lecturing. Teaching English abroad. Researchers at the government research laboratory, Department of Health, staff recruitment consultants and a major local authority. Trainee librarian. Neighbourhood Services Officer. RAF and Army officers. Personnel officer and management trainee with major telecommunications companies. Management consultant.

Manufacturing, utilities and production

Process controller with United Utilities. Product specialist with engineering company. Business planner with Scottish foods company. Supervisor with major British brewing company.

Finance

Trainee chartered accountants with most of the national and regional firms. Analysts with various building societies. IT officer with major life

---

insurance company. Claims assessor. Insurance underwriter with major European insurance company. Accounts officer with major high-street bank.

Sport, entertainment, travel and leisure

Travel agent. Professional county cricketer. Professional basketball player. Chef. Journalist. Crew duty officer with major UK airline. Customer services with train operating company.

Careers in Geography (see the Association of American Geographers' website at http://www.aag.org/careers/) asks the sensible question 'What can you do as a geographer?' In reply they list 133 careers which American geographers have gone into. The *GeoJobs* publications in New Zealand and Australia all give examples of how individual geographers have worked through the early years of a very wide range of careers.

In short, geographers can take all kinds of jobs in the public and private sectors, and this cushions them against the vagaries of recruitment in individual areas. The other valuable feature of geographers is that they can be both specialists and generalists. Many will specialise during their final year in technical areas of the subject and use those skills in their careers – examples include medical geography, hydrology, conservation, Geographic Information Systems and computing skills, or land management. Additionally, geography graduates are generalists – good communicators in writing and orally, numerate and computer literate – with a wide understanding of how the world works and how people influence the physical environment. We can play to both strengths as needed – high technical skills in specific areas and/or flexibility in what we can tackle.

That is why geography graduates have one of the highest rates of graduate employment.

A number of observers have agreed that geography as a subject has very useful features as a preparation for life in general as much as for a career. Here is what some commentators and employers have said, and the authors are very grateful to the Royal Geographical Society (with the Institute of British Geographers) for gathering these people's views.

Geography provides a liberal education that transcends traditional disciplinary boundaries in bringing together social and natural sciences – people and their environment. It deals with issues that are central for society and it equips the young for their future.
Professor Andrew Goudie, Pro-Vice-Chancellor, University of Oxford

Numeracy, literacy and geographicity are the things that people in underdeveloped countries need to help them with their decisions.
Baroness Chalker of Wallesey (Linda Chalker), UK Minister of State for Overseas Development (1989–96)

The understanding of geography is central to industry for the efficient delivery of goods and services, and the commercial sector has an increasing need to employ people who understand the interaction between people, environment and society.
Vanessa Lawrence, GIS Business Development Manager, Autodesk Ltd

The geographers we have recruited are well-organised, are able to structure their thoughts and actions most efficiently, and have very clear views of their career paths.
Pene Axtell, Recruitment and Training Manager, Carnaud Metalbox plc

A UK geography department asked its recent graduates what they particularly valued from their geography degrees. Here is what a cross-section of them said.

I really benefited from the applied skills I learned.

The research skills I got were very useful for my job in marketing.

The friendly treatment I got from the staff – very accessible – suited me perfectly.

The field trips and the brief time we had getting to know professionals/officials in the 'real world' were memorable and beneficial.

I am a property consultant and so analytical skills, data analysis and report writing were the things I got from the degree that have been invaluable for my job.

It increased my confidence to do things.

AT SOME PLACES CONTINENTAL OR WORLD PRICES ARE SET – PLACES OF POWER.

FIGURE 5 *The flower market at Aalsmeer, The Netherlands*

FIGURE 6 *The trading floor at the Kansas City Board of Trade, Kansas, USA*

I went on to an MPhil in Social Sciences and then into consultancy in social and market research, so the key elements were the range of courses, survey methods and the dissertation – that's what I now do daily.

Research strategy and time management and interpersonal skills based around the projects.

The variety of the course has helped me to talk to people better [she is now in business banking].

As a TV news researcher, I needed the all-round education and the broad-based degree.

Report-writing to deadlines [district manager for a brewery].

In my accident investigation work, the transport geography module was perfect for this job.

Perhaps the key point here is how you will come to value the many different parts of your geography degree, and often not the obviously 'geographical' bits. You may be unable to predict while at university which will turn out to be the crucial aspects for you in later life.

In Chapters 4 and 5 we shall show you how departments of geography are trying to achieve their educational goals and so prepare you for the types of working life you are likely to have in the first half of the twenty-first century. But for the moment you may like to undertake the following simple exercise.

ACTIVITY 9

Visit your school or university's careers office. Check what employers in your ideal job area are looking for when they advertise jobs. What sort of jobs are they advertising? How do you shape up for the skills they are asking for? Does the careers office run any courses on how to apply for jobs or how to do well at job interviews?

3.5 A GAP YEAR?

More students are choosing to take a gap year between the end of school and the start of their university studies. It is often 12–15 months long because the UK school year traditionally ends in June/July and the university year starts in September/October. The growing popularity of the gap year is partly a matter of fashion and keeping up with fellow students who are financially able to enjoy a period of world travel. In other cases a gap year may be a matter of necessity as you build up the savings from a job to see you through university. Whatever the motivation for the gap year, the question arises of how you can make that quite long period as useful as possible for your geography degree course and your subsequent career.

One obvious advantage of the gap year is the ability to travel to new places inside or outside the UK and such a widening of horizons is particularly valuable for a geographer. You will come to appreciate different cultures and that can help you put the British experience into context as just one of many ways of organising a society and economy. For the physical geographer the ability to observe different climates and landscapes at first hand will bring classroom teaching to life in very vivid ways.

Of course, travelling is not only an experience, it is also a test of how good you are at organising yourself and negotiating your way out of the difficulties that travel and new situations seem to throw up. You may learn some foreign languages and that is always good for your curriculum vitae/ résumé.

Travel is one focus for a gap year but not the only one. Another is to gain employment. This may be simply to earn the money to help you pay your way through university. Yet even this can be valuable if it also develops work-related skills like working in a team, dealing with the public, planning activities or training people. Try to analyse any gap-year jobs for their 'deeper' benefits to you as a person.

Of course, you may be able to get not just any job, but one that is actually in the sort of field where you hope to work. For example, the job may be with a conservation organisation (if an environmental career seems attractive) or with young people (if a teaching job is a possibility). This has the added value of providing you with front-line insights to the realities of working in that field, and potential employers are more likely to recognise your sincerity in applying for jobs with them.

Working to earn is one possibility; another is to work in the voluntary sector in the UK or overseas to help those in greater need. Giving something back to the community can be a very rewarding experience.

The final set of gap-year activities may be obtaining top-up qualifications. You could take an academic subject or acquire a generic skill by enrolling on an IT course, for example.

In practice, 12–15 months is a long time and therefore many gap-year students try to combine several of these things. So, six months travelling in Australia is funded partly by six months' paid work in an area of career-relevance in the UK before leaving, and in the final three months before university the days are used for voluntary work and an evening IT course is also taken. Perhaps this is a bit of a tall order but you get the idea.

There are now a large number of companies selling exciting travel and adventure holidays, some with a work element perhaps in a voluntary or teaching capacity. These come as well organised (and often expensive) packages – not much chance here to show your organisational skills. Unless, of course, you use the high cost as a reason to develop fund-raising skills. This can be a creative and enjoyable activity in its own right, and a proven track record in this should impress future employers.

A gap year can be a memorable time in your life, particularly if you plan it carefully to get the most out of it. You also need to consider your wider aptitudes so that the gap year suits you while still challenging you in the ways you want it to.

4 UNDERSTANDING THE LEARNING AND TEACHING OF GEOGRAPHY

Among the hundreds of geography departments around the world there is a wide range of courses and degree structures. However, some features are regularly found. Early in your degree you will probably follow a set of courses/modules common to all the students on your degree scheme. Later in your degree you will probably have some choice as to which courses you study. Most of the courses will be taught using a relatively small number of teaching methods. How geography courses are taught varies much less than the actual subject matter of the courses. So, you will usually be taught geography through:

- lectures;
- tutorials;
- seminars;
- fieldwork;
- practical classes; and
- a dissertation or project.

Additionally, you will be able to learn geography through the Internet and using other resources. Each is a distinctive way of introducing you to geography. Your role in the learning process varies from the apparently fairly passive (taking notes during a lecture) to the obviously highly active (as during your dissertation and fieldwork). This chapter explains how each of these methods of teaching and learning works and what you can do to get the most from them.

4.1 LECTURES

> *He who can, does; he who cannot, teaches* [he who cannot teach, lectures?].
> George Bernard Shaw

Here is some traditional advice to new staff on how to write a lecture:

> *First, you tell them what you are going to tell them (set the scene). Second, you tell them. Third, you tell them what you have just told them (recap and summarize).*
>
> *What I tell you three times is true.*
> Lewis Carroll

The lecture is probably the single most frequently used method of teaching geography (and most other subjects). A lecture involves a timetabled period of usually one hour at a regular time each week during which a member of staff will talk about some aspect of geography. The lecture topics will usually be listed in the syllabus section of the course's guide. With so much stress placed on lectures as a method of teaching, it is worthwhile reviewing why they still form the mainstay of teaching geography in higher education.

What are lectures for? Lecturers are trying to do one or more of the following things:

- start you off (and no more than that) on your study of an aspect of geography;
- give you key facts you need to know (e.g. dates, places, events, theories, formulae, data);
- give you an overview of the structure of a large field of geographical research, focusing on the essentials;

* show you how a geographer develops an argument;
* get you enthusiastic about the subject so you will want to study it further;
* challenge the *status quo* in a part of geography and suggest alternatives to the current orthodoxy;
* challenge you to re-think your views on a part of geography.

It is the last two of these – being critical of orthodoxy and of your own views – which are the most important. The critical thing is for you to be critical. Kneale (1999: 65–71) develops in more detail the idea of critical thinking in geography.

The lecture format allows an expert in a particular aspect of geography to give you an overview of the subject based on his/her extensive reading and perhaps research. That wealth of experience and understanding, distilled into a 60-minute presentation, allows you access to the key points of a large volume of work by geographers. As such, it should be a sound platform for you to begin your learning about that part of geography.

Lectures will probably not only be retrospective (in the sense of reviewing previous work) but also be forward looking, identifying the key issues for future policy or for our theoretical understanding of the subject. This programmatic aspect of the lecture will be most strongly developed in final-year courses and can be useful as a source of ideas for your dissertation or essays.

Given the popularity among staff of the lecture as a teaching device, it is perhaps surprising that the traditional lecture is criticised by lecturers as well as by students. If a lecture is to hold the class's attention for 60 minutes, it needs to be delivered in an enthusiastic manner. A lecture is, to an extent, a performance and the students are the audience; not all lecturers are top-rank performers. So some lectures will be rather dull. Also, some lecture topics are important but hard to convey in an exciting way. So, one of the criticisms of the lecture is that it is actually quite hard to concentrate on a subject for as long as 60 minutes. Studies have shown that student attention is high at the start of lectures, declines slowly to a low point after 20–30 minutes, where it remains until attention picks up again in the final 5–10 minutes. Sometimes lecturers will use various devices to counteract this cycle of attention. Some are quite simple, such as varying their tone of voice, moving to another part of the lecture theatre to talk, and showing slides or overhead transparencies (OHTs). Other ways of breaking up the lecture include asking you questions during the lecture or getting you to discuss a geography topic with your neighbour for a short period in the middle of the lecture.

Lectures are also criticised as being too lecturer-centred and hence too passive an experience. To an extent this is inevitable with this style of teaching; other teaching methods such as tutorials (which require you to be more involved) are used to counteract this. However, this criticism is also partly wrong. You should be active in lectures – you need to be thinking about what the lecturer is saying, and summarising the lecture in your lecture notes. This requires a lot of focused effort; it is just not as visible an effort as speaking in a tutorial or rushing around on fieldwork.

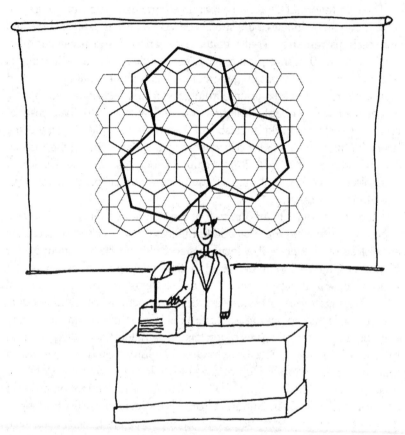

'Okay class, copy down quickly Christaller's k=132 system'

Here are a few tips on how to get the most out of your lectures.

1) If you cannot attend the lecture (perhaps because you are ill), copy someone else's notes. You will still learn something from a lecture at second hand.
2) The key things to look for in a lecture – and to record in your notes from it – are the *structure* given to the topic by the lecturer, the *key contents* (such as dates, definitions, people, formulae, events, data and theories), and the *key arguments* which are described.
3) You need to be able to write notes quickly and accurately. Why not try to develop your own 'shorthand' of abbreviations for the frequently used terms in your courses? Here are some examples – *U* for unemployment, *A* for Africa, *Gn* for glaciation, *env* for environment, *dev* for development, *ch* for change. This is not unlike text messaging on mobile 'phones.
4) Remember that the most important thing for you to do during a lecture is not to take notes all the time (though 60 minutes of feverish scribbling is a serious temptation) but rather to *listen actively*. A tape recorder may capture precisely all the lecturer's words but it will not tell you which ones are important and why they are important. Only you can do that by thinking about what you are hearing, sifting out the key issues and noting them down.
5) You may want to experiment with different methods of taking notes; for example, lots of headings and sub-headings rather than text and sentences; using parallel columns of notes for different sections of the lecture; or graphical methods like mind maps and spider diagrams where items get put in boxes on the page and the boxes are linked by lines to show how the ideas in the boxes are connected.
6) After the lecture, make sure your lecture notes are properly labelled in terms of the course name and the lecture title. In your first year at university you might attend as many as 200–250 lectures, so obviously some sort of filing system will be helpful.
7) Soon after each lecture read over your notes and make sure they make sense and that you have got down all the points legibly. If not, check with a fellow student at the next lecture to see if he/she can fill in the gaps in your notes. Don't re-write your notes.
8) Think about the topic; go over the key points in your mind. Why is this topic important enough to warrant a slot in this course? How does this lecture relate to the previous ones in the course? Are there

parallels between this lecture and those in other courses? Do you agree with the lecturer and what he/she said or the approaches or emphases used? A lecture is not a brainwashing exercise, no matter how eminent the lecturer is on this topic. You are expected to develop your own views and be able to support them.

9) Any lecture is only a summary of a huge volume of material. It is like a map, showing you the intellectual topography of a subject area. For the full detail that you will need if you are to do well in your essays and examinations, you will need to do the reading which has been recommended by your lecturer. Lecture notes are the starting point to develop an understanding of a subject – and so accurate comprehensive lecture notes are better than scrappy ones – but they are only the start.

10) After the lecture you will need to do the follow-up reading. It might be fun to share out the reading with a friend and then compare notes. Explaining to your friend the key points in what you have read will consolidate that material in your mind, and you can learn what was important in your friend's reading.

The traditional phrase 'reading for a degree' does actually describe what the ideal student should be doing – as a guideline, two or three hours of reading for each hour of lectures.

KEY TIPS

✓ **Don't make notes on everything the lecturer says – just the key points.**
✓ **After the lecture, read over your notes and jot down the most important things you have learned from the lecture.**

Taking notes

Notes are just that – the key points to remember. Notes must be much shorter than the original but be faithful to it. Taking good notes is a skill in its own right. This is easier if you are reading a book or article since you can check and re-read and get the points correctly, but the danger is that you copy down too much detail to remember it all. In a lecture you hear the information just once (though the good lecturer may help you by showing OHTs or PowerPoint displays of key facts, by giving you a paper handout with the key points or having a website you can check later). But even downloadable lecture notes on the web are not a substitute for the lecture itself.

Lectures are experiences as much as information-transfer devices and that experience and the lecturer's enthusiasm for the subject can help you to process and understand the topic more fully. But since you hear it once, you have time to note only the key points, and even that demands close attention – no easy matter for 50–60 minutes at a stretch. The spelling of new words, statistics or formulae can be hard to get right first time and may need subsequent checking. Notes for an essay on the topic will have to be more detailed than notes from a standard lecture.

Here is a block of text on dams. It might be from an article or the text of a lecture. Try to take notes from it while you are reading it over for the first time. To make this a realistic exercise (your time will always be limited) spend not more than 10 minutes on this – time yourself.

DAMS

Dams are one of those aspects of geography that transcend the human, environmental and physical divide in geography. They change natural physical systems in major ways. They have considerable environmental consequences. They are designed to achieve clear economic benefits. Dams also raise considerable passion, on both sides; their supporters and critics are equally vociferous. A dam will have long-term effects yet it must weather rapidly changing social, political and economic conditions. Finally dams are as much symbols as real features of the landscape. They can symbolise progress and people's mastery of the natural world, or they are emblems of our uncaring destruction of the environment and harsh treatment of the poor, depending on one's point of view.

There are about 45,000 large dams in the world today; that is one large dam opened per day since 1900. They cross 61 per cent of the world's large rivers and their reservoirs cover an area as large as Great Britain and hold as much water as the North Sea. The USA was a keen early dam builder. The USSR followed suit with an ambitious programme between the 1930s and 1970s, while many African, Asian and Latin American countries built dams from the 1960s. Dams now irrigate up to one-sixth of the world's food production and produce one-fifth of the world's electricity.

The economic case for dams is apparently simple. Once built, they provide cheap hydro-electricity and can also be used for many other economic activities, ranging from flood control and irrigation downstream to fishing, recreation and shipping services upstream. The profiles of costs and benefits of dams differ from those for conventional power stations. In the latter, the cost of building the station is lower but the running costs are higher than for dams. Dams require huge amounts of capital and take many years to build, in part because of the need to make dams enormously strong to resist the water pressures on them from the reservoir. Yet targets and reality may be some way apart. A quarter of all dams irrigate less than 35 per cent of the area planned and another quarter deliver less than half of the promised water to cities. Average construction costs of dams over-ran by 56 per cent.

The environmental arguments are much more complex. On the one hand, flood control may alter the natural environment, but in ways that help flood-prone human communities which are more important than any changes to wildlife or habitats which the dam may cause. Silting may be a serious problem (the reservoir traps the silt that would have been washed seaward, so reducing the reservoir's storage capacity and power-generating potential). This process reduces global reservoir capacity by about one per cent per year. The Nile Delta is in retreat because the river now has so much less silt. While a conventional power station can be demolished when its useful life is over, dams are very difficult to remove; in effect, they are quasi-permanent. A number of other environmental effects from dams have been cited. The weight of water may be great enough to cause earthquakes. The raised humidity along the reservoir's banks may cause insect-borne diseases to increase. While the hydro-electricity may be generated without any direct emissions of greenhouse gases, the rotting of the vegetation along the reservoir's banks and the intensified agriculture allowed by the regular supply of water, may release even more damaging amounts of methane. The lack of new silt downstream may mean that more artificial fertiliser is needed on the farms.

Socially, dams also have many consequences, and these often focus on the effects on the communities along the river. Downstream, irrigation water that was once free becomes something controlled by others and may be expensive to buy, though it is now available year-round and not *just in the wet season. Upstream, whole communities, habitats and archaeological sites may be lost for ever – 100,000 people were displaced by the High Aswan Dam – and the communities may never be compensated. Fishing communities tend to have their lives seriously affected by dams due to fish migrations being stopped or fish being killed by changes in water quality.*

Much depends on the balance of effects, and that will vary from river to river, country to country and between historical periods. In the USA, for example, money may be available to compensate those who lose out from the dam being built, and to remove archaeological sites and even to recreate habitats. In poorer countries there may be enough money to build the dam and no more than that. In arid areas the silting and health issues may be more of a problem in some river systems than in others. A lot also depends on how widely one casts one's boundary for measuring the effects of the dam. Is the boundary just the dam itself or are the indirect effects on the wider regional and national economy to be taken into account too? Is one looking at effects in the short term or the long term? One's judgement on the fairness of damming rivers may depend on whether the beneficiaries adequately (if at all) compensate the losers.

Finally, how does one weigh up the relative importance in the balance sheet of issues as diverse as electricity prices, national electricity self-sufficiency, rare species, a traditional way of life and a loss of power among local communities? Much depends on the values of the person compiling the balance sheet. There may therefore be serious arguments over the likely advantages and disadvantages of building a new dam that will go beyond the ever-present difficulties of making forecasts (particularly long-range ones) and scientific uncertainty. Such controversies have been witnessed during the last 20 years over several proposals for dams to be built in China and among those to be funded by the World Bank in several countries. Overall, the effects of dams are elusive for any one river/dam, and are very hard to generalise for dams worldwide. All in all, dams are a real microcosm of many of the issues which face geographers today.

Your notes

Your notes on this 1,018-word text might look like this.

DAMS

controversial – econ., soc. & env. effects
D as symbols of modernity or destructiveness
45K big D in world. 61% large rivers D'ed; up to 1/6 world food irrigated by D;
1/5 world elec. from D

Econ. case
cheap renewable elec.; multi-use; capital intensive but cheap to run; long term; BUT may deliver < planned irrig./water; 56% cost over-runs

Env. issues
Stop floods (good for river people);
habitats + sites lost
trap silt (not to fields – problem); ↓ power (–1%/yr); coast (erosion – Nile Δ);
health effects (insect diseases); ↑ methane (rotting)

Soc. issues
way of life changed
people displaced (100K High Aswan) – go where?
balance of power from local to outsiders?
fishing ↓

Effects vary
- *by river, country, period*
- *LDCs not afford remedial measures*
- *neg. effects local/quick but benefits regional/national, longer? Any local compensation?*
- *understanding of effects ↑;*
- *effects' balance depends on perspective + world-view + self-interest;*
- *precise measurement/costings v. diff.*

D as microcosm of geog.

These notes are 165 words long, just 16 per cent of the length of the original which was packed with detail. Some notes could be under 10 per cent of the original and still keep all the key points. See how contractions have been used (for example, 'D' for dams) and symbols (↓ for 'reduction in'). Observe also the use of headings, punctuation and gaps which show the basic structure (such as 'Econ. case', 'Env. issues'). This is much clearer to read than a solid block of text (as in the box below) which is exactly the same 165 words but laid out without spacing. It is much harder to grasp.

<u>DAMS</u> controversial – econ., soc. & env. effects. D as symbols of modernity or destructiveness. 45K big D in world. 61% large rivers D'ed; up to 1/6 world food irrigated by D; 1/5 world elec. from D. Econ. case – cheap renewable elec.; multi-use; capital intensive but cheap to run; long term; BUT may deliver < planned irrig./water; 56% cost over-runs. Env. issues – Stop floods (good for river people); habitats + sites lost, trap silt (not to fields – problem); ↓ power (–1%/yr); coast (erosion – Nile Δ); health effects (insect diseases); ↑ methane (rotting). Soc. issues – way of life changed, people displaced (100K High Aswan) – go where? balance of power from local to outsiders? fishing ↓. Effects vary – by river, country, period, LDCs not afford remedial measures, neg. effects local/quick but benefits regional/national, longer? Any local compensation? – understanding of effects ↑; effects' balance depends on perspective + world-view + self-interest; precise measurement/costings v. diff. D as microcosm of geog.

Why, you might ask, take any notes? Why not just read the original, learn shorthand and so lose none of the detail, or use a highlighter pen to mark up a photocopy? The main benefit of actively taking notes is that this forces you to read the text carefully. You have to work out what is important and see the sequence of the argument. You are actively working with the text and that helps you understand the issues. Details can always be found again if you need a particular fact, statistic or quotation for an essay.

PITFALL 6 EQUIFINALITY

Large-scale farming (for example, ranches and plantations) can be the result of:

* colonialism (as in East Africa);
* the settlement of areas previously uninhabited by white people (Australia, Canada);
* the amalgamation of previously smaller farms (see East Anglia or northern France);
* socialist economic planning (e.g., in the former Soviet Union).

Different processes can lead to similar outcomes. Put another way, you cannot necessarily infer the causes of events solely on the basis of their end-result.

4.2 TUTORIALS

You may also be taught geography through the medium of tutorials at some point during your degree. A tutorial is a small group of students (outside Oxford and Cambridge, usually 6–12 students) who meet with a member of staff for an hour, often weekly or fortnightly. Tutorials can have two functions, one pastoral and the other academic.

The pastoral function

Some tutorials have a pastoral function – they are a device to allow you to discuss academic and personal problems with a member of staff who may be able to sort them out. Your department can use the tutorials to keep a watching brief on your progress. The general rule is that unless a problem shows clear signs of going away of its own accord, it is better to tackle it quickly. The earlier problems are dealt with, the better; problems often become harder to solve the worse they have become.

Remember too that your university will probably also have other sympathetic knowledgeable people whose skills lie in helping with difficult issues on a confidential basis – examples include your Students' Union, a student-run Nightline, Chaplaincy, Student Counselling, a Learning

Support Unit or a college tutor. It would be a terrible waste if financial, medical or personal problems got to such a state that they threatened your continuing at university. For many people university may be their first time away from home, so some problems can be expected. Tutors will often have good ideas from their experience of previous students about how you can tackle any problems.

Academic functions

The other function of tutorials is to help you learn more effectively about geography and to develop new skills – why else would departments continue to use so expensive a method of teaching? Lectures are a much more 'efficient' way of teaching a large number of students.

Staff would probably argue that the benefits of tutorials derive from the way they can help you. These are:

- acquiring critical judgement (learning how to assess the strengths of various positions and arguments);
- active learning (you can be asked to learn geography in many different ways during a tutorial);
- practising how to apply principles to cases (beyond the examples given in the lectures);
- challenging attitudes and beliefs (higher education is a chance to think afresh about ideas);
- developing oral skills (you and 200 other students in a lecture theatre cannot hold a debate, but in groups of 6–12 you can);
- gaining practical skills (some of which are better taught in smaller groups);
- generating self-confidence (in your growing abilities);
- learning from other students (listening to what they say and how they argue and work);
- learning to work in a group (tutorials are just the right size for small-group work);
- promoting understanding (through debate and having the time to think ideas through);
- taking more charge of your learning and reflecting on your progress.

What happens in tutorials

Group work

A tutorial can be used to let you practise working within a small group of students to achieve together some common goal. You might be asked to produce a group report on some aspect of geography – together you discuss it, share the reading and research, and all contribute to the final report. Generally essays are completed by each student individually – you personally are solely responsible for it and the quality of the final essay is attributable to you. In a group project, however, you are usually collectively responsible. Somehow the group has to decide what to do, which is no easy task if, say, three vocal students each want to develop the project in a different direction. So, in a group, everyone has to learn:

- to weigh up the pros and cons of different tactics and find the best one;
- to compromise with others so that a single plan of action can be agreed;
- to work fully within that plan (even if there are parts of it that you do not much like) so that the final report will be as good as possible.

There is a general model of how groups form and work which is useful to bear in mind. The model (its originator was B. W. Tuckman) predicts that most groups go through four stages in their development.

Forming	*The group members are polite, a little wary of each other, finding out how each other works and reacts.*
Storming	*People start to dislike each other and discover each other's agendas and shortcomings; some hostility may be evident; the group could collapse at this stage, or its members realise that they may have to compromise with each other.*
Norming	*Realisation that compromises have to be made, so ground rules are worked out to get the job done despite each other's weaknesses and using each other's strengths.*
Performing	*The conflicts are worked through and a modus vivendi is established; people get to know and even like each other; and the job starts to get done to a common plan which everyone accepts.*

The moral of this model of group work is that tensions are inevitable; time has to be allowed for them to be worked through. So don't expect to get productive work done too quickly, not until the group has started to gel. Of course, the second project your group does will get going much quicker than the first.

Group work has its practical problems. One such problem is the lazy student who skives off and hopes to benefit from everyone else's efforts without contributing anything him/herself. The rest of the group will be expected by the department to try to persuade the lazy student to join in. If you cannot, tell your tutor so that the lazy student does not get a free ride on your efforts. The opposite problem is the bully – the pushy, loud-mouthed student who drives the project in his/her direction ignoring all others. If everyone agrees that the proposed direction is the best one, then that is fine. If, however, the rest of the group has good grounds for disagreeing, you need to use your force of numbers to persuade him/her to back down. Again, the necessity of compromise is critical.

Working collaboratively is rarely straightforward but the experience of working together through the difficulties to achieve a successful outcome can be among the most rewarding and valuable experiences you have at university.

'The joys of group work'

A useful device to help the group work well is to give every member a specific task. So, one person might agree to chair the group, a second might take charge of word-processing the final report, a third might draft section A, and so on. Each task is essential, big and important, and they all contribute to the overall progress of the project. For further details about group work there is good advice in Vujakovic et al. (1994).

An effective team is likely to be one which:

- agrees a plan of action and sticks to it . . .
- . . . yet can be flexible when difficulties arise;
- trusts each other to work well and is not disappointed;
- helps each other out in a crisis;
- is sensitive to each other's needs and uses each person's talents in the best way;
- learns quickly from its mistakes;
- reviews progress regularly;
- provides positive outcomes for all its members.

Group work in a tutorial is as much about learning how to work reasonably harmoniously and effectively with fellow students (and practising this skill) as it is about the specific topic of your project. Of course, in your career after university you will also have to work in groups, and so the experience will be useful and should be recorded on your curriculum vitae or résumé for potential employers to see.

ACTIVITY 10

After your next group project, jot down how well the group worked as a team.

What went well? Why? What did the group do to produce this?

What went less well? Why? What did the group do to produce this?

What did each team member get out of the project?

Could you list a few do's and dont's to help achieve an even better outcome the next time you have to work in a group?

It is worth going through this list after each group project.

Discussions

The other main function of tutorials is to practise discussing geography. This lets you learn new material that comes up in discussion. It will also help you understand things that puzzled you or were not clear in lectures or your reading – you can ask the tutor questions so he/she can explain things more fully. But the main skill you will learn is how to discuss issues sensibly. Essays and examinations are about improving your *written com-munication skills*; tutorials are about improving your *oral communication skills*. This is partly a matter of gaining some self-confidence in speaking and you will never develop that if you spend the entire tutorial staring fixedly at your knees, hoping that by avoiding eye contact with your tutor you will never be asked to speak. If there is a debate, say a few words. If someone says something you do not fully agree with, say something like, 'I see what you mean, but what about . . .?', or 'how does that fit with . . .?', or 'will that be true of every region?'

If the tutor asks you to prepare a 10-minute talk on something, the best preparation is to practise your talk several times beforehand so you can almost memorise it. If you have a script of what you want to say, try to lay it out with headings and sub-headings in big clear handwriting or typeface. Giving the group a handout to study while you are talking can also be effective. This handout might include photocopies of relevant maps and diagrams on your topic. You can use the handout for detailed facts and the talk itself for the overall structure and key ideas.

Remember, no one is going to laugh at you because everyone else is probably going to give the same kind of talk and will not want an aggressive tone applied to theirs.

Finally, your tutor may be the person who marks your essays, especially in Year 1. So he/she is in an ideal position to give you constructive feed-back, directly on your essays and more indirectly on your general progress. If you feel that the feedback is not detailed enough to be really useful, don't be afraid to ask for elaboration. After all, it is only through con-structive criticism that you can build on the things you are already quite good at and improve where you are rather weaker.

So, although tutorials are an expensive way to teach you, they do have clear aims that many geography departments value highly enough to justify continuing to teach in this way.

You will learn a lot from tutorials but only if you participate in them fully.

KEY TIPS

In a tutorial:

✓ **Listen and learn.**
✓ **Speak and contribute to the work of the group.**
✓ **Speak early on, so are a part of the process.**
✓ **Enjoy working with staff and student colleagues.**

PITFALL 7 I → N DIMENSIONS

'Edinburgh is a tourist centre.'

Of course, but Edinburgh is also:
a capital city, cultural centre, financial centre, university city, historic place, a city with slums, poverty, drugs and HIV, a port, etc.

Images are often one-dimensional; they miss out a lot about a place. But despite their partial (even caricatural) nature, they are often very influential for people's behaviour. Is that true of the process by which you chose or are choosing your university?

4.3 SEMINARS

'When I use a word,' Humpty Dumpty said in a rather scornful tone,
'it means just what I choose it to mean – neither more nor less'.
Lewis Carroll

A seminar can take various forms but it usually means that a large group of students (perhaps the whole class or a third or half of them) meets for an hour or two to listen to one or more students giving a presentation on some aspect of a course. You may already have been asked to give a short talk (say, 10 minutes long) in the informal setting of a Year 1 tutorial. That will be good practice if you have to give a seminar which will be a more formal presentation, usually in the later years of your degree. It will be more formal in the sense that it may last longer (say, 15–30 minutes), you may have to present the talk from the front of the class (as the lecturer would) and you may be expected to use presentational aids such as a slide projector, overhead projector or handouts. In the past every student might have given an individual seminar presentation. Ever-rising student numbers may mean that today you will be part of a group of students giving a group presentation. So the question is this: how are you going to use the nerve-racking business of a formal presentation to maximum advantage? Remember, your presentation may be assessed by the tutor and/or the class, so it is important to do it well.

PITFALL 8 *Ad hominem*

'Smith was opposed to birth control because he was a Catholic.' It may be that Smith's Catholic faith predisposed him to this position, but he may also have had other reasons for being against restraining population growth. These other arguments also need discussion. Try to avoid personalising (that is, *ad hominem*) arguments in essays and in seminars.

To give a good talk you need to progress stage by stage.

1) As you would for an essay, analyse the question or topic. Pick out the key ideas, events or approaches to it – your background reading will help here.
2) Write out fully what you want to say, highlighting the key points. But do not read out your text; talk to your audience. Remember that this talk is to be listened to, not read, so keep your sentences shorter than you would in an essay. Also, listeners cannot go back and re-read – they hear it once only. So help them by having a clear structure and giving them audible clues (e.g. 'First, I want to talk about . . .; then I shall move on to . . .'). Remember also to recap, so that points which your listeners may have missed can be picked up.
3) A talk will often be helped by visual material. This could be a handout that summarises the headings and sub-headings of your talk, and shows the audience complex or detailed material such as maps, graphs, formulae, dates and statistics. A collage of material can be produced for

a handout using the scale-reducing facility of a photocopier and scissors-and-paste. You may well find the raw material for this handout in the books or articles you read or on the Internet.

This material might also be presented to the audience as overhead transparencies (OHTs) or 'foils' – your tutor can advise you on the different sorts of transparencies and the photocopiers locally which will make OHTs. They are a particularly effective means of showing the audience the main features of your talk, using headings and sub-headings that you can talk to. They can also act as a prompt for you, the speaker, so you are not tied to your script line by line and have something to fall back on if you 'dry up' for a moment. Just make sure they are legible – a minimum of 20 point typeface and not too many words crammed on to each OHT.

If you have access to a reliable computer display system, then a PowerPoint presentation can be impressive, but have a set of OHTs as back-up in case the system crashes – it often does!

'How not to use an overhead projector'

Giving the talk is tricky; at the front of the class, you are rather 'on stage'. Experienced speakers would give you this advice.

1) Practise the talk several times so that you have confidence that you have enough material (but not too much) for your allotted period of time.

2) Perhaps get a friend to listen to one of your practice sessions. Does your friend think that you spoke too slowly (boring) or too quickly (a gabble that was hard to follow)? If so, adjust your speed of delivery and perhaps the amount of material you are trying to cover in the time.

3) Speak a little more slowly than normal and build in pauses so people can take notes. Repeat key points and summarise what you have said.

4) Try to look up and talk to the audience as much as you can, as opposed to reading from your notes all the time. The use of OHTs helps here (but don't stand between the overhead projector and the screen, so blocking the audience's view!).

If your talk is as a member of a team, your group will need to meet regularly to plan the whole talk, divide it into sections and allocate these to speakers. However, a single handout is still needed and each speaker should 'hand over' the rostrum smoothly ('. . . and now I hand over to Sam who will talk about . . .').

If your presentation is going to be marked, the criteria usually employed will include: audibility, clarity, structure, use of visual aids and handouts, and interest. So bear these issues in mind as you prepare your talk. Further details about how talks are assessed are given in Section 5.7.

When the seminar presentation is over, reflect on the fact that the next one you give will be a little easier as you build up experience and confidence. Being able to give a coherent and interesting talk (whatever the subject) is a useful skill for job interviews and later life generally, which is why geography departments make you give seminars. So, having gained the experience, remember to note it in your curriculum vitae or résumé as another proven skill to your credit. Further advice on giving talks can be found in Hay (1994) and Young (1998).

KEY TIPS

✓ **Practise your talk beforehand.**
✓ **Visual aids will help you and your audience.**
✓ **Talk to your audience; don't read out your script.**

4.4 LEARNING WITH THE INTERNET

The Internet – the global network of interlinked computers – is potentially a very valuable resource because it can help you in four ways. First, you can use the Internet to send e-mail messages. E-mail can keep you in touch with:

* your tutor (particularly useful if he/she is hard to get hold of or if you are a distance-learning student);
* fellow students, perhaps including those working on group projects with you;
* and perhaps your family and friends.

E-mail also lets you send whole documents over the Internet (called 'attachments') as well as short messages. So, you might be able to submit your essays by e-mail rather than by post or handing them in personally. It is well worth getting familiar with how to use your local e-mail system – your department or computer service provider will show you the current procedures.

Secondly, it is possible to join discussions over the Internet using online conferences, tutorials and discussion groups, perhaps with members from all over the world. These can be a very useful way of joining the debates on key issues and getting a fresh perspective on geography.

Thirdly, your university will use the Internet to provide you with information about your courses, the library and the various departments. You can use the local system to check what is in the library catalogue, to recall books out on loan or check the university's rules and regulations. Some courses will have their lectures and other material on the web so that you can use these anytime.

Finally, the Internet is useful as a source of information – a global library which never closes, but is so vast it can be very hard to find what you want.

Using the web critically

The web has many advantages as a resource for studying geography because it makes a vast 'library' of study materials available to you wherever you are and whichever type of computer and modern software you are using.

Unlike a library, the web does not shut down in the evenings. Its contents are never unavailable because someone else has borrowed them. The web can give you access to resources even if your library is small and even if the resources you want are specialised (those from a foreign country, for example). Most of the time the web gives you quick access (particularly with the fast machines and connections found in universities) though morning access in the UK tends to be faster than that in the afternoons when North American users have come online. The web is also huge. The amount of material available from the web is still expanding very rapidly indeed.

However, one's enthusiasm for the web also needs to be tempered because not everything is on the web. Most academic books are still available only in paper form – how else is the publisher to make a profit other than by selling books? On the other hand, because publishers receive subscriptions from universities, online versions of some journals are becoming more readily available. Statistical material is less well represented on the web than on paper and much historical source material can be consulted only in its original paper form or on microfilm/microfiche.

Finding appropriate material on the web can be problematic. Whereas libraries are traditionally well catalogued by author, subject and title so you can rapidly find items, the web has grown up in a quasi-anarchic fashion. Search engines, available through most PCs, allow you to search the web for key words but these do have the tendency to produce thousands or millions of 'hits'; certainly far too many for you to inspect to see if they are what you need.

You can try to narrow down the search process by using more precise terms. So if you wanted to research urban tourism, searching for *urban* AND *tourism* together (the AND all in upper case) would give a more manageable set of results than either term searched for separately. Better still may be to search for "*urban tourism*" (the words being together in double quotation marks), which should bring forward only those items which contain that exact phrase. Also, rather than search for *urban* or *city*, search for a specific city (say, *London* or *Paris*) and then follow up links from that site. Using an asterisk in your search term (e.g. *Americ**) will bring up all materials relating to America, American and Americans, which is useful if you are not quite sure of the exact search term you need to use. As a rule of thumb, if what you want is not in the first 50 items presented to you by the search engine, it is unlikely that you are going to find it through that search. Better to try another search. Avoid – or at least be aware of – ambiguous search terms like 'ford' (the river crossing, vehicle manufacturer or ex-US president?), 'port' (a harbour, the drink or nautical term) or 'corn' (wheat, maize or painful foot condition). Some web users believe that the Google search engine (http://www.google.com) is better than others.

You may know the exact name of the organisation you want to contact. Most web addresses use the organisation's name or acronym (e.g. BBC) followed by a code for the type of organisation (.ac for universities; .com or .co for businesses; .gov for public authorities; and .org for non-profit-making groups). The address ends with a country code such as .uk, .fr, .ca, .de (Germany) and .ie (Ireland). US addresses uniquely have no country code and use .edu for universities. Using this pattern you can often guess correctly an organisation's web address.

Another way to find what you want is to use a portal. This is just a website where a wide range of individual web addresses that may be useful to a particular group has been gathered into one place. Here are some useful portals for geographers.

- The CTI Centre for Geography at Leicester has a very good set of links for most branches of the subject
  http://www.geog.le.ac.uk/cti/info.html

- Fundamentals of Physical Geography
  http://www.geog.ouc.bc.ca/physgeog/links/links.html

- Site linked to P. Kneale's book, *Study Skills for Geography Students: a Practical Guide* (1999)
  http://www.geog.leeds.ac.uk/staff/p.kneale/skillbook.html

- CYBERGEO – a French site but in English
  http://www.cybergeo.presse.fr/revgeo2.htm

- The 'Geography for the New Undergraduate' project at Liverpool Hope University
  http://www.livhope.ac.uk/gnu/

- GEOsources is a Canadian site with a good range of source materials
  http://www.ccge.org/geosources/Jumpstn.htm

- The Social Sciences Gateway gives you access to a very wide range of useful material.
  http://www.sosig.ac.uk/

In Appendix D we provide a list, correct as of early 2002, of websites which we feel can be useful for geography undergraduates. Bookmark the websites that you find useful.

Caution

You need to be aware that websites can disappear or be radically altered at any time. So, websites (including all those referred to in this book) will sooner or later cease to operate. Books, in contrast, may go out of print but they are never 'unpublished'; they still exist and can be read.

Referencing websites

If you want to refer to material on a website in an essay, it is important to give the correct reference to where you found the information. The reference for a website should have three parts:

- *the real name of the site, including its provider's name (which is what appears in your text as the cross-reference to the full reference at the end of the essay);*
- *the full address/URL of the actual files or pages you used;*
- *the date when you visited the site since, unlike paper publications, websites can be changed after publication.*

Here is an example.

> *In the text of the essay you might write this:*
> *'Hurricane Zebra caused 1,000 deaths in Ruritania in January 2002 (BBC News Online, 2002)'*

> *In the References section at the end of the essay you give the full website reference:*
> *BBC News Online (2002) http:// www.bbc.co.uk/news/(the additional filenames to this story).html (30 March 2002)*

So, the web gives you excellent access to material, but is that material any good? Is it reliable for your studies as a geographer? The books in your university library have all been checked by the publisher and recommended by the geography staff because they are known to be worth reading. With the web, however, no one may have checked the material for its truth, authenticity or reliability. At worst it may be wrong, out of date, biased or even totally misleading. It is virtually the case that anyone can put anything on the web. So you need to check web material for its validity. Here are some useful checks.

1) Is the author of the material a reliable, trustworthy body or person?
2) Even such reliable bodies may give a slanted view on the issue, so what might that slant be, and which other bodies might give the issue a different slant? Try to find their websites to get a balanced view of the evidence and arguments.
3) Are there references to other work and independently checked research backing up what is said? This does not guarantee that the website is fair but it may help you spot highly contentious sites.
4) As you read web material check if the presentation seems to be giving you the whole story or only a careful selection of words, images, evidence and argument.

Of course, these checks and this alertness to what you are reading should be something you display with everything you read (books and articles as well), but it needs to be kept particularly in mind with web material.

A useful website to help you to evaluate other websites (!) can be found at *The Internet Detective* (http://sosig.ac.uk/desire/internet-detective.html).

The future university?

If you put together all the facilities of the Internet you could create a virtual degree course using the web (and a few already exist). You read lectures on the web, join online tutorial discussions, search the web for information, submit essays over the Internet and even go on virtual field courses to distant places without ever leaving your computer. All of these aspects of the Internet already exist individually and many are now routine.

ACTIVITY II

If you can get access to the Internet, find a website (say, that of your own town, school or university) and see how accurate, up to date and fair it really is.

1 Is it clear, interesting and easy to use, or is it a jumble of information?

2 Is it telling current and potential visitors or students what they need to know or is it just public relations and hype?

3 What does the site contain and what is missing?

4 Is it fully up to date?

5 Does it give interesting links to other Internet sites?

4.5 FIELDWORK

Work expands to fill the time available for its completion.
C. Northcote Parkinson ('Parkinson's Law')

Fieldwork is fun (even when it is done in the pouring rain). It is one of the distinctive features of a geography degree. Yet it inevitably takes up a lot of staff and student time; organising it safely is demanding; and it can place considerable financial burdens on students and departments. In general, students are now required to do less fieldwork during their geography degrees than previously. Some fieldwork, formerly conducted during the classic week-long residential field course away from the university, is now being replaced by day excursions in the university's local area. There are even some early attempts at virtual field courses during which you never leave your computer yet 'travel' to distant places. These rather lack one of the important unspoken merits of residential field courses – the way they let staff and students get to know each other well – which is often cited as one of the reasons for the generally good staff-student relations you find in geography departments.

Many will argue that fieldwork is central to geography. It shows you how places work and how they differ from each other. It lets you practise investigating the real world and it focuses attention on the way economic, social and physical processes are integrated and interact in particular places. Fieldwork, perhaps inevitably focusing on relatively small field sites, emphasises the smaller scale of geographical processes (people, businesses, local organisations) and the way (inter)national forces affect small areas, and how small areas react to these external forces.

Yet within geography there are debates on the role of fieldwork today. Some human geographers will argue that fieldwork is unnecessary since, like other social sciences, human geography is concerned with processes and theoretical approaches which have little need for real-world verification. Geography should be an intellectual training and not a practical one. Fieldwork, they might argue, concentrates too much on the unique and the specific to the detriment of our understanding of general spatial processes. This group would seek to minimise the fieldwork component of geography degrees, hence saving on staff time. Other human geographers and many physical geographers would still subscribe to the traditional justification for fieldwork given earlier in this section, particularly when it is carried out in distinctive and unfamiliar environments. They might also argue that fieldwork needs to be better integrated into the

EACH GENERATION LEAVES A DISTINCTIVE IMPRINT ON THE LANDSCAPE.

FIGURE 7 *Neolithic village at Skara Brae, Orkney Islands, Scotland*

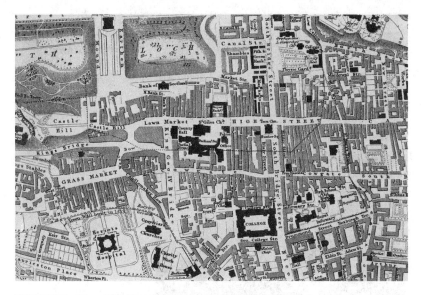

FIGURE 8 *The Old Town, Edinburgh, Scotland*

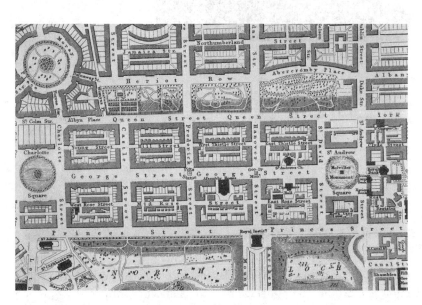

FIGURE 9 *The New Town, Edinburgh, Scotland*

FIGURE 10
Motorway junction, Perth, Scotland

'Rain'

...and to your left you will see a splendid example of an erratic.....sorry about the rain and fog.....

curriculum in terms of project design and skills acquisition than it is currently. So the role of fieldwork is the subject of some debate among geography staff.

Whatever the balance of opinion on fieldwork in your department, the main things to try to gain from fieldwork (on day excursions and residential courses) are these:

- learning about places;
- learning specific skills;
- learning how to structure a field project;
- discovering how to work around practical difficulties;
- appreciating the limitations of field methods.

These are your tutors' objectives when they take you on field visits and courses.

Learning about places

Fieldwork is fun because it takes you to new places and gives you the chance to study them in detail. You get to see why that place is distinctive, how it is changing and 'what makes it tick'. It teaches you how to observe and be curious about places and that is how research often starts – observation and asking questions.

Learning specific skills

A field excursion can be used to teach you how to use a specific method of research or piece of equipment which cannot be demonstrated in the classroom. You have to go out into the real world to learn how to conduct a questionnaire survey of the public in the street or how to measure the speed of a river's flow. The range of skills you may be taught in the field is wide – for example, mapping, surveying, sketching, identifying plant communities, sampling soils, measuring landforms, surveys and interviews of the public or officials. You may not know which of these will be useful to you in the future, so note carefully how they are carried out and any pointers to good practice in their use. Even if none turns out to be directly useful, you have at least demonstrated your ability to learn practical skills and that is a skill in its own right which is worth having.

Learning how to structure a field project

Most days, the work on a field course will be structured as you would a small research project, with a number of phases.

- Set out the aims, research problem or research questions for the day; all research, including your dissertation, needs to be clear about what it is trying to achieve.
- Select the appropriate method to tackle this research problem. Every method has its strengths and weaknesses, so choosing the best one requires an appreciation of the various methods and a reasoned argument as to which is best under particular circumstances.
- Implement the field methods – actually using the methods or equipment in the real world to collect the data or information that you need to meet your aims.
- Collate and analyse the information you have collected so as to make sense of it and see what light it sheds on the original research problem.
- Write a concise report describing the four stages above and your conclusions, illustrated as appropriate by maps, graphs or photographs (see Lewis and Mills (1995) for more detailed advice).

If these stages have not been clearly set out in a handout or briefing session, ask the tutor to explain what you are being asked to do.

Other research-type projects, and particularly your dissertation, will probably follow the same five stages, only on a bigger scale. So each day's work on a field course is an example in miniature of how to plan a project. In that sense it is worth looking behind the immediate detail of the work to see how the lecturer has constructed each task.

Discovering how to work around practical difficulties

Fieldwork is about practicalities. Things will go wrong from time to time: the tide is too high for the beach survey; your interviewee is not in when you call; the rain prevents your photography; the equipment you were to use breaks down. Fieldwork should be faultless but rarely ever is. Learning how to work round such problems is a key skill in effective fieldwork. Employers like practical people who can show that they can cope with difficulties.

Appreciating the limitations of field methods

All field methods are good for some things and worse for others. They work better in some circumstances than in others. There are better and poorer ways of putting them into practice. A good field worker appreciates these points and so is able to choose the best methods for any particular research situation. That is a useful skill to learn while on field courses. **Notes on safety during fieldwork are given in Section 4.11.**

So to summarise, fieldwork is a surprisingly 'deep' experience. There is the surface level of exploring a new place – travelling somewhere, carrying out an investigation and finding out why that place is different from others and how it works. Then there is the deeper level of acquiring new research skills and learning how to structure a piece of research anywhere in the real world. Finally, there is the level of critical appreciation of both specific research methods and of the use of case studies and field-work: you learn to assess how far they can improve our understanding of geography. Fieldwork has its limitations; you need to have done some fieldwork to really appreciate them.

PITFALL 9 DETAIL AND SCALE

Below is the course of the River Lune, north of Kirkby Lonsdale, Lancashire, as shown on the Ordnance Survey maps at the scales of: (a) 1:625,000; (b) 1:250,000; (c) 1:50,000.

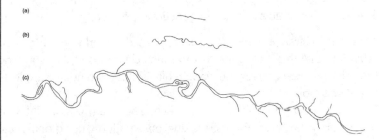

(a)

(b)

(c)

FIGURE 4 *'The River Lune'*

The bigger the scale of the map, the more detail you get. The detail may lead you to understand better the process that created the feature. Maps are deliberately-constructed versions of reality for their scale, not reality itself. Similarly, social surveys conceal behind their averages and totals the details of the wide range of conditions among the individuals who comprise the survey.

KEY TIP

When in the field, ask yourself:

✓ **What makes this place different?**
✓ **How has this place changed?**
✓ **Why is this place changing?**

4.6 PRACTICAL CLASSES

In many geography courses/modules, you may be required to undertake 'practicals'. Practicals can be found right across geography but they are most prevalent in physical geography and the teaching of methods of geographical research (e.g. statistics, cartography, computer-based methods and GIS). They are usually based in a laboratory of some kind (rather than a lecture theatre or seminar room) and they may last for 1–3 hours. You may be able to do some practicals in your own time; others may require staff to be present to instruct or for safety reasons. The practical exercise may be given to you as a handout, in a work-book or online and you will be expected to work through each exercise.

There are four main reasons why practicals are used.

1) They can illustrate a theoretical concept in a real-world situation. You often find that this makes the theory much clearer as well as showing how diverse and complex the world really is; it is often not as simple as in the textbooks.

2) Practicals show you how to do research. They train you in defining problems, testing hypotheses, making observations, using analytical techniques and equipment safely and accurately, and presenting the results clearly and concisely.

3) Practicals train you in specific skills (such as designing a questionnaire or measuring the speed of flow of a river) which are key skills in parts of geography.

4) Finally, practicals can be fun, breaking down any barriers among students and between students and staff.

There is a very wide range of activities that can benefit from some practical application. In physical geography you might analyse the composition of soil or water samples. You might study maps to examine the links between the physical environment and land use. You might use computers to practise the statistical analysis of data, simulation modelling or analysing text qualitatively. The computer might give you access to the Internet which might be the source of information you quarry. In human geography practical exercises might involve creating a video documentary, practising interviewing skills on fellow students, compiling questionnaires or working on historical documents. Some of the most imaginative types of teaching can be found in the practical exercises staff devise to help you appreciate geography.

Practicals are just that – sessions to teach you how to do practical things. You need to understand what you are being asked to do, why a given procedure is useful and, in a critical sense, what its strengths and weaknesses are. Above all, you are learning how to do something carefully, precisely and successfully. Practical exercises are more likely to have a single right answer than essays or examinations. Usually practicals are assessed by writing up the aims, methods and results in a concise report with data, output or diagrams attached. Section 5.5 gives you more details on how to write up a notebook.

In addition, practicals are useful not just for a particular course, but also as a way of building up a set of skills that will come together again in your dissertation – the ultimate geography practical. Getting the most out of practicals really involves regular attendance, attention to detail, clear notes, a crisp write-up and a critical approach that leads you to appreciate when a practical skill should or should not be used and, when it is, what its limitations are.

Practicals may be done by a group of students rather than individually so the ideas on group work in Section 4.2 are useful.

Notes on safety during practical classes are given in Section 4.11.

Learning new skills is important. Learning how to learn a new skill is even more important.

PITFALL 10 FALSE PRECISION

'38.347962% of the sample had taken a foreign holiday in the last 12 months.' 'The mean river speed was 10.786351 kph.' Your calculator or computer probably calculated those figures correctly, but an answer to six decimal places is usually excessive – a spurious level of precision given the limited sensitivity of the equipment and survey methods you were using. '38.35%' or '10.79kph' should be adequate precision for most purposes.

4.7 DISSERTATIONS AND PROJECTS

If you steal from one author, it's plagiarism; if you steal from many, it's research.
Wilson Mizner

In most geography departments you are likely to have the option of producing a project or dissertation. Often it is a compulsory part of the geography degree, which shows you how highly geography staff value the dissertation.

The terminology varies but a project might be a research-based piece of work 5–7,000 words long, while a dissertation will be longer, perhaps 10–15,000 words long. Both will be the equivalent (in terms of your final degree classification) of 1 or 2 full units of study. Dissertations are often started in your penultimate year at university and handed in sometime during your final year. The summer vacation between the last two years can often be used to carry out field research for your dissertation. Many geography staff would view the dissertation as the most important single element of the geography degree, because it is here that you bring together all the skills and intellectual maturity you have acquired. The dissertation is your chance to show that you are a good all-round independent geographer. Many departments place considerable impor-tance on the dissertation and expect you to put a lot into it. It is your chance to work on something that really interests you, and develop the study in the way you want it to go.

Fortunately, given its importance, you will probably be given specific training in how to produce a dissertation. There are also several very good books to help you with your dissertation which will repay study (Bell, 1993; Parsons and Knight, 1995; Burkill and Burley, 1996; Flowerdew and Martin, 1997; Kitchin and Tate, 2000).

A dissertation is as much a process as a product. The product is the final long report and the process is how you create that report. It is on the process that we shall focus in this section.

Choosing a topic

Probably the biggest decision comes right at the start: What are you going to study? What will your topic be?

- It has to be interesting to you.
- It should offer you scope to be creative and show the examiners your skills.
- It must be feasible for you to undertake.
- It must be safe to do.

Where might you look for ideas on your dissertation topic? Clearly a list of past geography dissertations in your department might be a useful source of ideas, though the poorer of these may not be very good role models! Your lecture courses may well have thrown up ideas about the key areas in geography today. Are there issues in the media you could explore? Another tack is to consider your possible career and then devise a project that would let you work in that sector or with that sort of employer. Alternatively, you could choose first where you want to do the research (your home area, perhaps) and then focus on what seem like the most interesting current issues there. However you generate the ideas for your dissertation, it is always useful to write them down and show them to a tutor for a quick reaction.

The research proposal

The next stage is to expand your preferred dissertation topic into a linked set of stages. The key questions now are: What? Why? And how?

- *What* are you going to study?
- *Why* is it important to study this?
- *How* are you going to study it?

The answer to the 'what' question will eventually become Chapter 1 (the Introduction) of your dissertation. The 'why' question's answer will become Chapter 2 (the Background to your study) and the answer to the 'how' question will become Chapter 3 (your Methodology).

Planning something as large as a dissertation is complicated. You need to be organised so you get everything finished on time. Therefore work backwards from the submission deadline, week by week, to fit in all the activities. You need to make plans that are feasible for you. Do you have the time, resources, skills, equipment and transport to do all that you want? How could you acquire these facilities? Is the topic safe and, if there are risks, how could you reduce them to an acceptable level? Finally, envisage the things that might go wrong – your computer breaks down, a key interviewee is unco-operative – and sketch out contingency plans for coping with these, as far as you can.

Literature review

Reviewing the literature is designed to set your dissertation in its academic and policy context. It sets out where research and understanding of a topic have reached, and what the unanswered questions are, one of which you will tackle in your dissertation.

Once you have defined a very broad area in which you wish to do your dissertation, it is advisable to read widely around the subject to see what the key questions are. It is often helpful to envisage your dissertation as being at the point of overlap between several big sets of literature. So, for example, if your dissertation was going to be about farming and pollution, then the relevant sets of literature would include the following:

- EU and national legislation on and approaches to pollution;
- trends in agriculture;
- agricultural pollution specifically;
- policy evaluation.

You will not need to read everything on each of these large topics (you would not have the time to, anyway). Only read those aspects which will affect your farm pollution topic. The advice on reading and taking notes in Sections 4.1 and 4.10 will be useful here, as is the advice on essay writing in Section 5.4.

As you read around your subject, always keep your topic in mind. You are not reading for its own sake; you are reading specifically to find material that will be relevant to your research, that can include ideas, models, data, references, examples, techniques or paradoxes. You are not so much reading as hunting through the literature, and you have thought enough about the topic already to be able to recognise your prey when you come across it.

During your reading you need to look behind the actual literature and consider what theoretical stance it is taking and what principles or positions it is based on. Does it take a free-market or communitarian position? Is it concerned with sustainability? Does it support a particular theoretical position or approach to the topic that you may already have come across during taught modules? Is it clearly process-based, quantitative, qualitative, mathematical or experimental in approach? Are there moral dimensions or practical consequences explicit or implicit in what you are reading? You may want to follow or question these underlying assumptions about how your topic should be approached; you will certainly benefit from being aware of them.

It is important to look widely for items relevant to your dissertation, utilising the World Wide Web, GeoBase, library catalogues, review articles (e.g. in *Progress in Human/Physical Geography*) and the latest journals and books on your topic. Whenever you find something useful, remember to note carefully its full reference as well as the interesting material itself.

Research design

Research design is about specifics and so it is particularly difficult to give general advice on it. However, there are four issues worth raising – theory, topic, practicalities and argument. Your chosen methodology will have to be consistent with, and appropriate for, all of these.

1) *Your theoretical stance*   This will have some effect on your methodology. If you are a positivist and realist who admires the 'scientific method', then you will tend towards types of enquiry and methods (field measurement or experimental/laboratory studies) which produce statistical data and which emphasise sampling and inference, replicability of results, simple explanatory models and the identification of general trends or processes. If your theoretical position is more idealist or social constructivist, then you will use different methods designed to achieve a better understanding of how individuals interact with each other and perceive and understand their world.

Here are some questions which may help you to define your theoretical position.

- How objective are you hoping to be?
- How representative of wider conditions do you want your results to be?

* Do you want to fit your work into a wider structure?
* Do you want the reader to empathise with the people studied?

2) *Your topic* Your topic will also affect your methods. A focus on land use, political struggle or pollution will lead to different types of research question and you will seek to find or use certain sorts of data specific to that topic. A study of agricultural land use in Ireland will need different types of data (and different ways of collecting it) than a study of environmentalism among new-age travellers in France. It is 'horses for courses' when it comes to research design and the reading you have already done on the topic may suggest some research designs and methods you could use.

3) *Practicalities* Can you find the people or data you want? Is it possible to travel to the field sites or to obtain the equipment needed to carry out certain research procedures? Can you learn the key skills required (for example, to operate a piece of equipment or learn a foreign language)? These issues will constrain what is feasible given the time and other resources available to you.

4) *Your argument* Finally your style of argumentation will also affect your research design. If you want to compare two areas or sets of firms or social groups, then your research will have to be structured in sections to cover these areas, sets or groups. If you want to compare those who have participated in a scheme with those who have not, the survey method will have to be able to identify and study these two groups separately. Look at the design below.

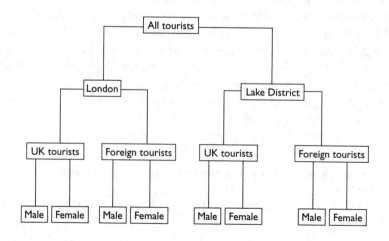

By carrying out sub-surveys of tourists in two places (London and the Lake District) you can compare different sorts of tourist experience (metropolitan and rural), which you cannot do if you confined your research to one case-study area. By separately studying UK and foreign tourists in each area you get a better view of which perceptions of place and tourist attitudes are common to all tourists and which are specific to particular nationalities in particular places. The gender division not only allows you to compare the general experience of being a tourist for men and women, but also lets you see whether that difference varies by nationality and for urban and rural holidays. A design as complex as the one above gives you a great deal of scope to tell an interesting story about tourism. It also requires you, when planning the study, to think ahead to the kinds of arguments and the structure of the results chapter long before you actually start the fieldwork.

The complexity of the research design above should probably be paralleled by a diversity of study methods. A study of tourism in London and the Lake District may need a general survey of large samples of UK and foreign tourists to find out what they think and do. It may also be helped by a more detailed study with a few tourists in both areas to appreciate their attitudes to these areas in more depth than a fleeting on-street survey can hope to give. You would probably want to interview key officials dealing with tourism and perhaps gauge the residents' views on the industry. A study of old maps and photographs and the media output about the area could also prove useful, as might an examination of the area's tourist image as promoted in brochures, guidebooks, photographs and souvenirs. In short, you should use a variety of types of survey and enquiry, each able to provide you with a distinctive slant or type of information and understanding on the topic.

PITFALL 11 ECOLOGICAL FALLACY

If regions with a lot of well-educated people have many cases of burglary, this does not mean that the well-educated are burglars. Features which are associated with each other at one spatial scale (here, educational attainment and burglary at a regional scale) are not necessarily associated with each other at any other scale (well-educated people as individuals being burglars).

In 2001 the World Bank showed that low-income countries which traded a lot had grown faster economically during the 1990s than low-income countries which traded less. From this result you might conclude that countries benefit more from free trade (a key feature of globalisation) than from protecting their home markets or aiming for self-sufficiency. This does not mean that all (or even a majority) of the individuals or regions within a country can expect to benefit from free trade. Whether that is true can only be determined by studying the trade/benefit relationship separately at the level of individuals or regions.

Different explanations and interpretations are often needed for different geographical scales.

The concluding chapter

This is perhaps the most difficult chapter to write and should not be rushed. You need to leave enough time to draft this chapter and then review it a few days later. The concluding chapter is not just a summary of the preceding chapters. Rather it is:

- a considered review of the key questions you set out early in the dissertation;
- an assessment of the effectiveness of your methodology for answering those questions;
- telling the reader what you have learned on the basis of your research results.

Avoid just repeating your results, survey by survey. Focus on the key questions, such as:

- what happened?
- why did it happen?
- which policies worked and which did not?
- how is the process (human or physical) working?
- is the situation (human or physical) improving or worsening in any sense?
- who has gained or lost out?
- what is affecting what?

This chapter is also the place where you might set out your recommendations for future policy or comments on the future course of events. These recommendations should be based on your results and not just on your personal views.

Some examiners may regard the quality of the concluding chapter as indicative of the overall quality of the whole dissertation.

Summary of advice

The keys to a successful dissertation are really quite straightforward – it is just tricky to remember and implement them all:

- a good topic with lots of potential is essential;
- tell your examiners which general geographical issue your dissertation is a case study of;
- be bold in your planning (it may not all come off but aiming too low at the start is far worse);
- examiners are often impressed when a student approaches a topic from different scales and uses different approaches;
- tell your examiners what you have learned from completing your dissertation;
- be boringly organised (it's quicker in the long run) since a dissertation is really as much a planning and management exercise as it is an intellectual one;
- expect something to go wrong, and don't panic when it does;
- keep in touch with your supervisor (he/she will help keep you on the right lines and up to the right standard, and his/her advice will be vital if there is a crisis);
- *and finally remember to enjoy your dissertation – it is the main part of your degree where you really are in charge and can develop along your own lines.*

More details on how dissertations are assessed are given in Section 5.6 and Appendix F. Notes on safety and risk assessment during dissertations are given in Section 4.11. Your ethical responsibilities while carrying out research are discussed in Section 4.13.

During your dissertation you will inevitably learn a lot about organisation, problem solving, independence and, indeed, about yourself. These lessons can be used to enrich your curriculum vitae or résumé. Details on this are given in Section 6.8.

KEY TIPS

✓ **At last, you're in charge of your studying!**
✓ **Be bold in your planning and careful in your work.**
✓ **Be clear on the aim of your work.**
✓ **What is the 'big issue' to which your dissertation will relate?**

4.8 OTHER METHODS OF TEACHING GEOGRAPHY

So far, we have looked at the principal ways in which geography is taught at university. In addition to these, there are other methods which are used less often.

Work-based learning

You might be given a period of work experience (sometimes called work-based learning) where you work outside the university with an employer or organisation on a project of mutual interest. Usually the department will set up the link and the project, and then train you in how to work with the employer. Work-based learning can be used to give you an appreciation of how real-world organisations operate and it may also require an end-of-placement report which will be assessed. Clear oral communication and good team-work will be important for the success of a placement, as will meeting deadlines and being a congenial colleague. Work-based learning can take many forms but you need to ensure that you record on your curriculum vitae or résumé what you did and what you gained from it. At the end of the placement you can prove to potential employers that you can work successfully in the real world.

Reports and the media

You may also be set various other types of exercise – writing a newspaper report in the style of a journalist on a geographical issue; producing a poster (see Vujakovic (1995) for advice); producing a video (see Lee and Stuart (1997) for guidance); or writing a guidebook to an area. Each of these exercises is giving you practice in different styles of writing for different audiences, and in Section 5.8 we expand on how to do well in these tasks.

Foreign exchanges

A few departments will have foreign exchanges where you spend part of your degree studying geography at a university in another country. Exchanges between the UK, North America and Europe are the most common examples. Aside from practical matters – the cost and possible foreign-language requirements – perhaps the key advantage with foreign exchanges is the way they let you experience another national culture and another university system. They are well worth considering, if available.

4.9 LEARNING TO USE OTHER RESOURCES

Among the other resources you can use to learn geography are the following:

* newspapers
* printed study packs
* computer-based modules and computer-assisted learning (CAL)
* collections of slides and photographs
* maps
* audio/video cassettes
* material on CD-ROM (such as statistics)
* microfilm/microfiche (for example, historical records).

Newspapers

Newspapers can be a useful source of information for the geographer interested in events and their portrayal. Newspapers used to be available for study only in paper form. Most libraries had the space to store only a few newspaper titles (perhaps one or two major national ones and some local ones). The arrival of microfilm and then microfiche copies of the newspapers from the 1960s reduced the storage problem, but 35mm roll film and microfiche are awkward media to use. Finding material was always difficult because, unless you knew the date an event happened or was reported, it was laborious to search through the newspapers to find all the material relevant to your study. In the UK only *The Times* had a complete printed index to articles; for other newspapers, their index, if they had one, was selective and limited.

This is one area where technology has transformed the usefulness of the source material. In the 1990s came CD-ROM versions of newspapers. These take up even less space so university libraries can increase the number of titles they take. The CD-ROM format also allows automated searching. You can type in some key words and all the articles on your topic will be listed for you to read. National newspapers are more likely to be available in CD-ROM format than local ones. However, the CD-ROM format often omits the advertisements and other minor elements of the newspaper and you cannot see the original layout. Photographs, graphs and diagrams are often not included either.

The advent of the Internet has allowed newspapers to put their editions on the web. This gives you rapid, user-friendly access to the latest issues. You can access a very wide range of newspapers from across the world (you are not limited to those your university library can afford to buy) and search facilities for key words are usually available.

Newspapers are useful in two senses. First, they provide sources of information of types not available in official statistics or textbooks – current affairs and local events, for example. Secondly, they are interesting for how they portray the news – indeed, for what they consider to be news-worthy. This in itself is a fascinating study. You can compare how a story is covered (if at all) by left-wing and right-wing newspapers, by local and national ones, by the broadsheets and tabloids, and by those in your country and in foreign ones.

Study packs

You will usually be guided towards the study packs and computer-based modules by staff at the appropriate point during your course. The collec-tions of slides and maps and the other resources are more likely to be useful during projects. Some of this material may be kept in your department, and some of it will be in the university library. There is more to geography than words in textbooks and journals, and these other resources can expand how you study places and communicate your findings.

Study packs and computer-based modules have the advantage that you can access them when convenient for you and work through them at your own pace. This can be particularly helpful for part-time students and those with other commitments. The chance to find your own materials to support projects is good training in being creative and learning how to find what you need – both of these are good career skills as well as enjoyable in themselves.

4.10 THE LIBRARY AND ICT

> *The true University of these days is a collection of books.*
> Thomas Carlyle

The library and information and communications technologies (ICT or computer systems) will be important resources for you throughout your geography degree. These are where you will find the references to supplement your lecture notes and the background reading for your essays. They are complementary sources. Libraries tend to be well catalogued (it is easy to find out what is in the library) but it may be difficult to get hold of the actual item when you want it, if it is in heavy demand. The contrast is with ICT. The information resources available through ICT are not well catalogued (it is not easy to find relevant, good-quality material for essays on the web) but once you do find something you want, it is usually easy to get a copy by downloading it (providing the computer system has not crashed!). So you will need to use both your library and ICT facilities to get the reading material you need.

Libraries – finding things

Each university differs in how it organises its library services: one central library or several departmental libraries; different classification systems for the books; different opening hours and borrowing arrangements. Yet there are a few general tips for using university libraries to best effect, bearing in mind that they are big, complex and heavily used facilities.

1) As soon as you arrive at university, learn how the library works. Pick up leaflets which describe the layout of the library buildings, their opening hours, and the length of time you can borrow different types of material. Walk round the library to get its layout clear in your mind. Practise using the online computer catalogue to track down books and practise finding them using the 'classmark' as your signpost to where they are shelved. Attend any training sessions on the library run by the university or your department. The library probably has web pages that will tell you a lot about how it works.

2) If a book or article is given as an essay reference it is likely to be in heavy demand. The library have may put it in a special collection with short loan periods. It is very useful to do the reading well in advance of the deadline for submitting the essay, even if you leave writing the essay until nearer the deadline.

3) If the library does not have the item you want (or someone else has borrowed it) you can use the catalogue to do a 'subject search' to see what else they have on that topic. Browsing along the shelves at the classmark where the missing item should have been may also help you find other relevant material on the topic. The library catalogue is often available through its website so you may not have to be in the library building to check its stock.

4) When you borrow books from the library, make sure you return them on time so that others can use them.

5) If you get stuck, ask the library staff for help; they are there to answer your queries.

ACTIVITY 12

Go to your university library and check that you know how to do the following:

- use the computer catalogue to check whether the library has a book or journal on the reading list of one of your courses;
- find the item on the shelves;
- borrow it.

You will need these skills throughout your degree course, so it will save you a lot of time if you learn them as soon as you can.

Having found the items you need to read, what next?

How to read (an academic article or book chapter in 20 minutes)

At university you will be given long reading lists for each module and will be expected to read widely on that subject. So, having found the books and articles in your university library, how do you read them quickly enough to get the work done on time and still have a social life?

There is not enough time to read carefully every word of every reference you are given. The faster you can do the reading, the more you can read in a given amount of time. Speed here does not mean speed reading. It means reading at normal rates but reading selectively so you find the key points quickly and make good notes on them.

- *Think why* you are reading the article or chapter, and what you expect to learn from it. What does the title of the article or chapter tell you?
- *Carefully read* the abstract/summary, the introduction to the article or the first paragraph of the chapter – this will give you the background and aims. (5 minutes)
- *Skim read* the other sections (literature, structure, methods, the rest of the chapter), noting examples or what seem to be key facts, arguments or examples. Often the maps, diagrams or graphs will be immediately helpful. (5 minutes)
- *Carefully read* the conclusions/discussion or final paragraph of the chapter. This will summarise the key results and the author's conclusions and interpretation of the issues. (5 minutes)
- *Make brief notes* (<100 words) on what you have learned from the article or chapter. This will summarise the message in your own words, which is useful for essays and examinations. (5 minutes)

So, 20 minutes to read an article and extract the key points.

Some of the tips on layout and contractions given in Section 4.1 on taking notes during lectures may be helpful here too.

ICT

As with the library, ICT services are organised differently in each university and these arrangements often change every few years as computing equipment, networking facilities and software are updated. The first step is to get your computer username and password when you enter university. These unlock the other facilities which are available via the computer system.

1) As with libraries, find out how the computer system works and what is available. So, pick up leaflets, attend training sessions and, armed with your computer username and password, log into a computer and explore the system.

2) If you bring your own computer to university, it is helpful if its software is compatible with the university's system. If you don't have your own computer, find out quickly where the public-access computer laboratories are and their opening hours (some will have 24-hour access).

3) You may already know how to word-process, but if not, learn as quickly as possible, preferably using the software which is standard in your university system. This will help enormously with your essays. Look for courses in your department or elsewhere in the university or use online training courses. Or even learn with a friend.

4) ICT may also help you get access to background material for your essays, perhaps using the Internet. You can do searches to see what is available on computers around the world on 'development', 'Africa' or 'glaciation'. The answer is likely to be that there is a huge amount available and that much of it is either irrelevant, too basic or produced by biased sources. It is difficult to find good relevant material via the Internet for university-level essays. If you do manage to find something, remember to mention in your essay the source of this information, using the web page's address (technically called its Uniform Resource Locator or URL) – see Section 4.4 for the details. Keep a notebook of the URL addresses of useful websites or 'bookmark' them if you have your own computer.

5) Use your ICT facilities to the full. Explore the available software and try it out. Some familiarity with a wide range of software will be useful in career terms.

6) If you get stuck with the ICT facilities, remember to ask for help from computer staff or fellow students; that is how to learn more about the systems.

ACTIVITY 13

Which new ICT skills have you learned in the last six months? This could be a single skill (such as sending an e-mail or creating a spreadsheet) or the use of a new piece of computer software. Add these skills to your Personal Record.

Which new ICT skills would you like to try to learn in the next six months?

KEY TIP

✓ **For a good degree in geography you need to know things and you need to know how to find out new things. Libraries and ICT can help with both of these.**

4.11 SAFETY AND ASSESSING RISKS

Every geography department should have taken care to ensure that everything it asks you to do is fully safe, especially in terms of laboratory practical classes, fieldwork and off-site projects and dissertations. Absolute safety can never be guaranteed, but staff should have taken all reasonable precautions against all reasonably likely risks. This includes your safety with respect to accidents in the laboratory, natural hazards (for example, when working near rivers or cliffs) and dangers to your personal safety while in rural and urban environments. You could ask to see the risk assessment of your impending fieldwork. You and your tutor should work through a risk assessment of any fieldwork or projects before you start.

If you ever feel concerned about your safety or the precautions that have been taken or should have been taken to minimise risks, you should bring your concerns to the attention of your tutor, course leader or someone in authority in your department. Equally, the department has the right to expect you to behave sensibly when in potentially risky situations and to take all the precautions you were told to take as well as exercising normal commonsense.

Assessing risks

There are some safety risks in studying any subject which uses more than just lectures and tutorials. When staff ask you to take part in practical classes, particularly those in physical geography laboratories, they will have checked the tasks carefully to ensure that they are safe when you follow their instructions on what to do. Similarly, when staff take you out in the field they will have assessed the route and activities to avoid any dangers, again assuming that you follow instructions.

Of more concern is when you plan your own fieldwork or laboratory activities – perhaps for a project or your dissertation. Here you are setting the agenda and so only you can accurately assess the risks. Most departments will actually require you to fill in a Risk Assessment Form and discuss it with a member of staff before you begin the work. The aim is to ensure that you have thought through all the stages of your work, reviewed each step for dangers, and have done all that is feasible to minimise those dangers.

Total safety is as unattainable on fieldwork as it is in any other aspect of life, but you can ensure that all reasonable steps have been taken to avoid unnecessary risks. 'Reasonable steps' might include the following:

- checking all laboratory practices with staff before you start them;
- not working alone in a laboratory;
- ensuring that you have been trained formally in the use of all the equipment that you might need;
- leaving a written note of your itinerary for the day with someone else and reporting back to them periodically;
- wearing a life jacket/vest when near rivers, lakes or the sea;
- taking a mobile 'phone with you when on fieldwork;
- checking whether the area in which you will be working has adequate signal strength for your mobile 'phone system;
- being accompanied on fieldwork by a companion;
- not working in dangerous areas in terms of physical hazards (for example, snow, cliffs, rivers) or in danger of personal assault.

Your tutors will have been trained to use their experience to ensure that you don't miss any safety-critical aspects of your work.

It is also important that you and your tutor bear in mind that issues like your normal level of mobility and the condition of your health may lead to a different assessment of risks for you than for some other students. If you are prone to epilepsy or suffer from diabetes or a heart condition, one may take a slightly different view of the 'reasonable steps' that should be taken, even when your condition is fully under control and not an issue during normal classroom and university activities. Your tutors need to know such things if they are to advise you fully.

4.12 COMPLAINTS

> **Franklin**: *Have you ever thought, Headmaster, that your standards might perhaps be a little out of date?*
> **Headmaster**: *Of course they're out of date. Standards are always out of date. That is what makes them standards.*
> Alan Bennett

It is unlikely that you will have a serious complaint about how you have been treated by your department, but occasionally things do go wrong. Each department should have both a code of practice of how ideally it should treat students and a formally approved complaints procedure. Both these documents should be widely advertised on departmental notice-boards and in handbooks given to students. It is usually advisable to pursue complaints according to your local complaints procedure, once you have satisfied yourself that something really is amiss. Sometimes it is best to take the matter up initially with the member of staff most directly concerned – the course tutor, for example, if you feel that an essay has been marked unfairly. In other cases (for example, if you are being harassed) the head of department might be a better person to contact. You could ask your year representative to raise the issue at the next meeting of the staff–student committee or Board of Studies. Your Students' Union may be another source of advice.

The exact procedure will depend on the nature of the problem, but most departments now accept that occasionally things will not happen as they should and that it is in everyone's interests that complaints are dealt with promptly, fairly and according to an established procedure.

4.13 RESPONSIBILITIES

It is hard to avoid this section seeming very stern. This is unhelpful but 'responsibilities' are a serious matter. Going to university implies a set of *mutual* responsibilities. On the university's part, they have to teach you well, provide the necessary resources for your studies, and support you academically and personally while you learn your subject. They need to strive continually to update the education they provide.

Your parallel responsibilities are worth setting out here. You are responsible for your education and making the best use of the university's teaching facilities (including the staff) and support services. You share with the university the responsibility for ensuring your safety (see Section 4.11). You need to act fairly on plagiarism and collusion (see Section 5.4). Try to avoid the use of language which may offend others (sexist or racist language, for example). You may meet opinions with which you seriously disagree; and others may disagree fundamentally with you. That starting point for academic debate needs to be handled carefully and with respect for others' sensitivities otherwise grave hurt can ensue. You have responsibilities to your fellow students generally, supporting their learning (getting a degree is not a competitive sport where everyone loses except the one winner) and you should certainly not impede other students' studies or harass them. Abide by the rules of the library and computing service. Also you need to work fully with fellow students during group work.

When you do research (projects, fieldwork or your dissertation) you have important responsibilities to act ethically.

* Don't study people covertly; their informed consent is needed.
* Respect local cultures.
* Keep information you collect safe and confidential.
* Be particularly careful when dealing with vulnerable groups who may not be able to give their informed consent (the young and elderly, for example).

- Protect the physical environment by avoiding erosion or damaging plant communities or animals.
- Don't endanger or frighten others by your behaviour.

Recognising these responsibilities and living up to them will help everyone at university.

Geography has traditionally used a wider range of teaching methods than many other subjects and the range has tended to expand further over the last few years. This is a major strength of geography at university, since it will give you a more varied experience of higher education than if you had chosen any other subject. Nothing but lectures, essays and examinations, year after year, might get rather tedious. This chapter has tried to explain the methods of teaching and learning geography and suggest how you can get the most out of them.

Of course there is also the question of how, as a geography student, you will be assessed, and the next chapter describes the different methods of assessment used in geography departments, and how you can do well with them.

UNDERSTANDING HOW YOU WILL
BE ASSESSED IN GEOGRAPHY

Throughout your geography degree you will be assessed by staff (and sometimes by fellow students) to see, basically, how good a geographer you are and whether your performance is improving. Many departments will use the trend in your marks as an important indicator. A falling trend will set alarm bells ringing and staff will be particularly keen to help you get back on track.

The most common methods of assessment are:

- examinations;
- essays;
- field and laboratory notebooks;
- dissertations and projects;
- oral presentations;
- posters, web posters and press reports.

In this chapter we shall describe each of these methods of assessment, explain why lecturers use them, what they and you can gain from this assessment, and tell you the features of a good performance in each type of assessment. It helps to know what examiners are looking for!

But before that, we need to explain a couple of general points about university assessment. The first concerns the relationship between academic progression and marking, and the second concerns how the marking process works.

5.1 ACADEMIC PROGRESSION AND MARKING

There are two models of how academic progression will affect the way your work is marked. The first, the 'one-standard model', takes the quality of work staff would expect from a final-year student as the one academic standard against which the work of students in all years is judged. If your department uses this model, you will find, not unexpectedly, that students' marks will be rather low in the first year and, on average, will improve steadily up to the final year. These low marks can be disconcerting for new students used to the higher marking scales at school; it really does bring you down to earth with a bump.

In the second model, the 'rising-standards' model, there is a separate standard for each year of the degree – that standard being what staff can reasonably expect students in that year to achieve. These standards rise in successive years of the degree scheme. If your department is using the rising-standards model, the average mark for a class may change little from first to final year (assuming that the effectiveness of your teaching and learning roughly keeps pace with the rising expectations of the lecturers). That apparent lack of progress can be rather dispiriting. Some students may be able to improve faster and so out-perform their colleagues and achieve a rising-mark trend.

Departments may not tell you (and may not even have collectively thought about) which model of progression and marking they are using. Hence it is not surprising that some mixture of the two models is used in practice. First-year marks are usually lower on average than final-year marks – that is the 'one-standard' model in operation. This tends to be tempered to some extent by an acceptance among staff that first-year students will inevitably know less and be able to argue and write less well than final-year students. So first-year marks will tend to flatter you a little and you will have to do better next year if you want to get the same marks in your essays and examinations. Only in factual tests (where each question has a single correct answer) should you be able to achieve high marks as easily in each year with the same amount of effort.

YOU ARE LIKELY TO GET A BETTER MARK IF YOU HAVE SHOWN THAT YOU HAVE UNDERSTOOD THE COMPLEX 'SHAPES' OF THE HUMAN AND PHYSICAL WORLDS.

FIGURE 11   *Sculpture by Barbara Hepworth in her garden, St. Ives, Cornwall, England*

5.2 WHAT IS BEING MARKED?

What are the staff marking when they read the work you submit? Is a mark of 60 per cent twice as good as a mark of 30 per cent and how do you measure 30 and 60 per cent? In some types of assessment (such as practicals and factual tests) the system of marking is obvious. The test is divided into many small sections, each has a single correct answer, and you either get a section right (and so gain a mark) or you don't. Your final mark is just the sum of the section marks and you can check this yourself when you get your work back.

However, for many types of assessment, such as essays and examinations, there is no such easy way of calculating or checking your mark. The examiner will have in his/her mind a checklist of things to be looked for: certain factual material, attention to key theories or approaches, a certain quality of argument and writing, evidence of reading, and correct procedures for referencing what you have read. However, the final mark will be an overall judgement of the essay or examination and not a simple summation of marks for particular parts of the work. So, ten students might each get a mark of 55 per cent for their answers to the same essay title and yet each will have achieved that mark by a different 'route', that is, a different combination of good and bad points. What the marker should do is to write a commentary at the end of your essay, perhaps supplemented by a mark sheet, which sets out what was good and bad about your essay, where you gained and lost marks, and how you could have improved the work. Criticism, provided it is constructive criticism, should help you improve.

You might wonder whether you can trust the marker's judgement. Your department should have given you guidelines to its marking, telling you what it expects and what the marks of, say, 50, 60 and 70 per cent mean for all the main types of assessment. Most departments also have a system whereby a second member of staff will check the marks given by the first marker. In the UK there is always an 'external examiner' – a senior geographer from another university – who is there to ensure that your department keeps up its standards and marks consistently and fairly. The procedure for making a complaint about the way your work has been marked was discussed in Section 4.12.

A lot of university assessment is about 'academic judgements' and the way staff view their specialisms. So it is very useful for you to try to pick up cues and clues. What do Dr X and Professor Y really want in essays and examinations? There will be many common elements – the rest of this chapter tells you what these are – but other things may be more related to the personal preferences and beliefs of the staff. It is worth trying to work out what these are, based on their lectures and publications and their comments on your previous essays. To an extent, you can think of university assessment as a 'game' with semi-public rules, and your task is to play that game as well as you can.

5.3 EXAMINATIONS

In examinations those who do not wish to know ask questions of those who cannot tell.
Walter Raleigh (1923)

The typical British university examination comprises a timetabled period of 1–3 hours during which you have to write answers to questions you have not seen before on a course of study that you have just completed. There are many variations on this basic pattern: there may be from one to four questions to be answered; you may have a free choice of questions or some specific questions may be compulsory. Mostly, the questions require a short essay-type of answer but sometimes briefer factual answers or even a battery of multiple-choice questions may be used. You will always be told beforehand what the structure of the examination is going to be, and previous years' examination papers will often be available for consultation in your departmental or university library or online. Sometimes the class will be shown the examination paper before the examination day so you can all research your answers; more often the examination paper is 'unseen'.

Examinations are still a favoured method of assessment among staff because they have the following advantages:

- you can guarantee who wrote each answer; this is not always possible with essays where students can copy from books or the Internet ('plagiarism') or from each other ('collusion') (see Section 5.4);
- everyone has exactly the same amount of time to answer the questions;
- the examination ensures you revise widely across the subject even though you will answer only a few questions (an unseen examination paper tests your *breadth of knowledge* whereas an essay tests your *depth of understanding* of a small area);
- you need to know the subject area well and have an agile mind to construct an answer quickly, both of these qualities being respected among academics.

Of course, from your point of view, an examination can be a highly stressful experience with so much depending on how you perform on the day itself. The fairness of the examination may be less complete than it might seem since illness could hinder some people on the day, and people react in different ways to stress – some positively and others negatively.

It may be useful to record here the commonest reasons why some people under-perform during examinations. You can easily convert these reasons into a list of do's and don'ts for your own examinations.

1) *Leaving all the background reading for the course until the last minute or not doing any at all and relying on lecture notes alone* Examiners are specifically looking for a much wider understanding of the

topic than there was time to give in their lectures. Use other authors' works and reference them in your answer ('The work of Smith shows that . . .', 'Jones's theory was important because . . .', 'During the debate between Brown and Green . . .').

2) *Not answering the question* You must answer precisely the question set, and not the question you desperately wish the examiner had asked. So, if the question is 'Explain the growth of cities in Latin America since 1950', you must do just that. 'Explain' (and not just describe); 'the growth' (rather than just their current structure); 'of cities' (and not rural communities); 'in Latin America' (and not other continents except perhaps a passing reference to how Latin American cities are different from or similar to those in other areas); and 'since 1950' (and not earlier periods except perhaps a modest reference to differences in urban growth before and after 1950).

3) *Poor structure to the answer* A good examination answer, like a good lecture, will have a clear structure which is set out in the first paragraph of the answer ('First, I shall . . . Then I shall . . . Finally. . . .'). The answer should then follow this structure, each paragraph being a discussion of a separate item.

4) *Limited knowledge of the subject* You will be expected to demonstrate a thorough knowledge of all the processes, approaches, theories, policies or events that are relevant to your topic.

5) *Excessive factual detail* This is the opposite fault to the preceding one. You have memorised the factual detail (good!) but you don't construct an argument which is supported by those details. The facts are usually the item of secondary importance in a good examination answer. The primary elements are the key ideas and arguments and the structure of the answer.

6) *Not writing enough* For each one-hour question (and assuming that you have normal-sized handwriting) you should aim to write at least four sides of the standard examination book used in universities. That means writing quickly but still legibly. If you usually 'write' your essays directly on to a computer, you may need to practise quick legible handwriting. Remember to leave yourself enough time to answer all the questions; don't overrun on the answer to one question and eat into the time left for the next one. One of the commonest faults is running out of time, meaning that the final question you answer is your shortest and poorest one.

7) *Waffle* This is a favourite comment by examiners. It means that what you have written is either irrelevant, a set of unrelated points or is

vague, lacking in details and could almost have been written by any intelligent passer-by who had not even taken the course.

So, if the points above are the pitfalls to avoid, how are you to get better at writing examination answers? The best advice is to practise by trying to answer the questions on previous years' examination papers, assuming that the course has not changed since then! Particularly useful is *practising the first five minutes of an examination answer* when you sketch your answer plan. Once you have got that right (detailed, well-structured, comprehensive, focused on the question that has been set), then the remaining period of the answer time should be the rather easier task of actually writing it out, paragraph by paragraph, according to the plan and as quickly and legibly as you can.

Hay (1996a, 1996b, 1997) provides further advice on sitting examinations.

Some good examination answers are sketched out in Appendix E, which shows some of the ways in which you can use the general advice here to help with specific examination answers.

KEY TIPS

✓ **Revise the key points, not everything.**
✓ **Practise planning the answers to examination questions.**
✓ **Answer exactly the question set.**
✓ **Show you have understood the topic and read widely on it.**

ACTIVITY 14

List your most recent examination marks. See if you can get the individual question marks as well as the overall marks for whole papers – but not all universities will release these. What is the trend in your marks?

Reviewing the questions you answered and their marks, can you explain the trend? Can your tutor help explain your examination marks?

5.4 ESSAYS

Learn to write well, or not to write at all.
John Dryden

The essay is another traditional form of assessment at university. The procedure is usually for you to be given a set of topics from which you select one on which to write within a word limit, often between 1,500 and 3,000 words. You may be given a list of references to books and articles which the lecturer feels you should read to help you produce a better essay, or you may have to rely on references in the course handout. In either case it is always worthwhile browsing the shelves of the library or searching its computer catalogue to find other books on the same topic. An essay is a chance to read deeply about one small part of the course, unlike the breadth of coverage which is the aim of the examination. Furthermore, an essay is usually a test in constructing an argument rather than just a listing of facts or chronicling of events. Departments will also expect an essay to conform to a traditional style of academic writing, as used in journal articles, for example. Most departments will give you detailed guidelines on the current acceptable style for academic writing with its many curious conventions (see also Turabian, 1987).

The title of an essay usually contains a command word – an instruction as to what you should do. Common command words include 'describe', 'explain', 'assess', 'evaluate' and 'discuss'. It is important to be clear what these command words mean.

- *Describe* is the simplest command and most often found in Year 1 essays. It means 'what happened, where and when?' (e.g. 'Describe the formation of drumlins').
- *Explain* means that you have to tell the examiner how some state of affairs came about (e.g. 'Explain the phenomenon of global warming'). This may involve a sequence of events or processes and some set of arguments as to why things had to occur the way they did. There may also be competing or conflicting explanations which you should set out.
- *Assess* is often used in the context of measuring the effect that one thing has on another ('Assess the effect of transport costs on industrial location'). The implication is that transport costs have some effect on industrial location (the extent of that effect may vary among countries

or industrial sectors or historically) but that there are also other factors which are simultaneously influencing industrial location. Your job is to say how important transport costs are in relation to these other factors. If the question was 'assess the influence of geology on coastal erosion', you would have to compare the effects of the geology with the effects of, say, climate, wave action and tidal conditions.

- *Evaluate* may be similar to 'assess' but is often used in a more general sense of asking you to discuss the good and bad points of some theory or statement ('Evaluate Smith's theory of migration' or 'Evaluate the utility of chaos theory in geomorphology').

- *Discuss* is the least precise of the command words but also the most often used! It is nearest in meaning to 'evaluate', but will also have elements of 'explain' and 'describe'. Suppose that the question was 'Discuss the effects of modern agriculture on European landscapes and wildlife'. You need to start by describing what effects modern agriculture has had on different landscapes; then explain why agriculture has had these effects; and finally evaluate how important these effects have been ecologically, aesthetically and politically.

Whatever the task you have been set by the command word, a good plan is to follow these stages.

1) Think about the topic; brainstorm it. What might it involve? What possible issues or approaches are there? What should you be reading about? When you start reading around the topic, what would you expect the books to be talking about? Jot down your ideas. This phase, so often omitted, is very useful for sensitising you to the subject matter and bringing into play what you already know of the topic from the lectures and your background knowledge.

2) Well before the deadline for submitting the essay, read the references for it and anything else on the topic which you can find on the library or on the web. As you read, you are looking first of all for the ideas you noted down in Phase 1 above. Secondly, you are looking for new ideas that did not come to mind in Phase 1. Take notes of the elements from your reading that seem interesting and may be used in your essay. Always record where (that is, in which book or article) you found anything you might use in your essay ('the source of the reference'). Mills (1994), Hay (1997) and Kneale (1999: 146–52) give more details about referencing.

Plagiarism is taking other people's ideas and making out they are yours.
Don't do it! Give the reference.

You might need to re-think how you read so as to get it all done. It helps
to have thought about the topic *before* you read anything, so you can spot
what is significant or useful. It also helps to learn to skim-read. Don't read
every word on each page but skim quickly through paragraphs hunting for
useful ideas and material.

- First, read carefully the abstract, introduction and conclusions to get the
  gist of the text clear.
- Then skim through the main body of the text, picking out just the key
  points and noting them down. It is much quicker, and just as effective,
  to skim read an item twice as to read every word of it once. The key
  points you are looking for are these:

  - the author's arguments, standpoints and ways of thinking;
  - definitions and ideas;
  - events and evidence;
  - problems and issues;
  - hypotheses, theories and explanations;
  - questions and conclusions.

* Finally, when you get to the end of the item, pause and reflect on the key points you have understood from it. Why do you think the lecturer recommended you to read this item? What did he/she hope you would gain from it?

If you can speed up your reading by skimming the texts, you will have time to read more. But always set a limit to how long you let this reading phase last. No one can read everything and you will not be expected to. If you can afford to devote a week, say, to producing an essay, then you can set aside only two or three full days for the reading. Do as much as you can in that time and then move on to the next stage. Section 4.10 and Kneale (1999: 48–52) provide some more advice on fast and effective reading strategies.

3) Back to the thinking again. After you have read all you need or have time for, think again about the essay topic and try to sketch an answer to the question. Where is your essay going to begin? What issues should it include? What might the conclusion be? You may need an essay plan which sets out the broad sections of the essay (for example, introduction, early theories, recent developments, conclusions) and then the sections can be sub-divided further into sub-sections, each of which could be a paragraph. Some people find it useful to write down each of their ideas for the essay on a small piece of paper; they can then literally shuffle the ideas around until they get a sequence of points that makes sense and answers the question that was set. The essay will end with a 'Bibliography' (a list of all the items you read for it) and the 'References' (a list of the sources of the quotations or facts you used).

4) Now you need to write the essay, paragraph by paragraph. If you word-process the essay, you can correct errors and amend your text very quickly. Word-processing is a skill well worth acquiring soon if you are not already familiar with it. Diagrams or maps may be added to the essay if they will help you convey your ideas. If you incorporate into the essay facts or text from another source (e.g. from a book or the web), remember to give the full reference to where you found it. This will avoid the serious charge of 'plagiarism', that is, passing off other people's writing as your own. It is also a serious offence to write an essay with another student (the charge of 'collusion') so that your essays are very similar (unless of course you were told to work in groups). When you have finished the essay, leave it for a day or two and then return to it so you can polish the style, spelling and grammar; it is important to get these right. If the essay ends up rather

longer than the word limit you have been set, you can now edit it, cutting down or removing sentences or sections that are not really essential or could be summarised.

PITFALL 12 SPELLING

Do you want to really irritate your tutor? Try putting lots of spelling mistakes in your essays – it works every time! Common mis-spellings include the following:

- parallel;
- symmetric;
- occurrence and occurred;
- separate;
- liaise;
- environment;
- medicine;
- its (unless it is in a quotation, never write *it's*; and never ever write *its'*).

Use the spell-checker on your word-processor, but if you are a UK student beware in case it is set to US spellings, and *vice versa* if you are an American student. The spell-checker will not pick up your use of 'form' when you meant 'from', nor will it tell you if you have omitted the word 'not', which does rather change the meaning of the sentence. Careful proof-reading of essays is still needed as well as spell-checking.

The only problem with Phases 3 and 4 above is that they work well for some people, who thrive on plans and carrying them out, but they do not work for those whose ideas become clearer only when they have written things down. For this group the normal process of revising the first draft of the essay may be far more than a tidying-up operation. Reading the first draft might alter their ideas completely and lead to a full re-write. In such cases you have to start the essay well before the deadline for handing it in, so as to leave time for the re-writing.

When you get your essay back, look not only at the mark you get but also at the comments. Every essay can be improved, whether it gets 75 per cent (an excellent university mark) or 45 per cent (a poor mark). So read carefully the lecturer's comments and think about how you might have done it better. Try to learn from each essay so that the next one will get a higher mark. If the lecturer's comments are not full and clear enough to give you ideas on how to do better, do not be afraid to ask him/her for more feedback. You can also swop essays with fellow students to see if there are any features of their style, structure or use of material that you can adopt.

Another tricky aspect of essays is the problem of knowing what the percentage mark you get for the essay means. You need to remember that, although there are differences among staff in what they value in an essay, there are many common understandings among them as to the features of good and bad essays. The key is to appreciate that a percentage mark is not a simple measurement of quality in the sense that a person's age or height can be easily measured by a single number. Essays are marked according to how they match certain profiles of good, average or bad essays and then the percentage mark is applied to the essay as a final summary of quality. So, an essay getting 60 per cent (an average mark) is much better than one getting 30 per cent (a fail mark) but not necessarily precisely twice as good (in the sense that a person aged 60 is twice as old as someone of 30).

Here is a guide to what marks mean in British universities and the features of essays that will lead to these marks. Your department should provide you with its 'grade descriptors' for essays and the other types of assessment they give you. The exact matching of percentage marks and 'classes' may vary a little among universities but this guide will be helpful.

First class; marks over 70%; A grades
> Precisely answers the question; has understood all the key aspects of the question; uses material from independent reading; factually accurate; covers all the main arguments; can deal with areas of academic controversy; can explain complex issues clearly; can evaluate competing positions; some originality of argument and independent thought; sensible, balanced treatment of the issues; very well structured and written.

Upper second class; 60–69%; B grades
> Answers the question; has understood most of the aspects of the question; uses some material from independent reading; factually accurate; covers most of the arguments; tackles to some extent issues of debate in the literature; reasonable explanations of complex issues; aware of debates in the literature; well structured and written.

Lower second class; 50–59%; C grades

Answers some aspects of the question; reasonable structure; mostly uses material from the lectures and only one or two other sources; deals with factual and descriptive material better than issues, arguments, debate or theory; may be lacking in specific examples; may have unsupported statements; structure may not be clear; written style will be acceptable but weaker in terms of fluency, spelling and grammar.

Third class; 40–49%; Poor grades

Only a few aspects of the question are tackled; may contain substantial elements of irrelevant material; almost no theory or higher-level issues; tends to over-generalise; solely concerned with events and facts; errors of fact; very little structure to essay; written style very poor in terms of spelling, grammar and powers of expression; only a basic outline of the topic and basic ability to write about it.

Fail; marks under about 40%

Not an acceptable essay for someone who has taken the course; shows little understanding of the topic; does not answer the question; shows little appreciation of the question set; unstructured essay; serious flaws in written English; could almost have been written by a passer-by who had not taken the course.

In general an essay which squarely meets one of these 'templates' will get a mark in the middle of the range (e.g. 55% or 65%). One which is approaching the qualities of the next higher class of answer will get a mark at the top end of the range (e.g. 58%–59%, or 68%–69%). An essay which only just gets into the class will get a mark of 50%–51% or 60%–61%.

This guide to what university essay marks mean is only a guide; individual lecturers will differ to some extent in terms of the value they put on factual material or good written style. This marking guide refers to essays. The marking of examinations will be similar, as will that for dissertations, whereas the marking of practicals will tend to use different criteria and a wider range of marks.

Finally, you will probably be given a deadline for completing each essay. These deadlines are important and your mark may be reduced if the essay is handed in late. So start each essay early. Don't leave it to the last week before the deadline – the books you need to read may have been borrowed by others and you are vulnerable to computer systems breaking down.

Further advice and examples of good essay writing can be found in Fitzgerald (1994), Hay (1995, 1997), Creme and Lea (1997) and Barrass (1995).

KEY TIPS

✓ **For each essay, leave plenty of time to read, think, plan, write and revise. Start early!**

✓ **Learn from each essay – what worked well and what needs more work?**

ACTIVITY 15

Gather together your most recent essays or other assignment marks and review the marks and comments you received. Do the same comments keep reappearing as good or bad points on your essays? What are they?

Focusing on the weaker points, what can you do to overcome them in future assignments? If you are unsure of the way forward, can your tutor help you?

5.5 FIELD AND LABORATORY NOTEBOOKS

The notebook requires a very different style of writing from that of examinations and essays. A notebook – it may be a hardbound volume, a ring binder or a wallet file of loose sheets – is essentially a record of events. It records the project you did on the field course or the analyses or experiments you carried out in the physical-geography or computer laboratory. As such, the emphasis is on the clear recording of methods, events, circumstances, observations and results. The written style should be concise – possibly even in note form. You will probably wish to include graphic material such as a graph of results, a sketch of the apparatus, a map of the field area, photographs or illustrative brochures of the locality. Finally, depending on the circumstances, it may also be appropriate to include a section on the conclusions you draw from this field study or laboratory procedure and also to set out the limitations of the fieldwork or laboratory procedure and the analyses. Recommendations may be made for further studies on the topic which would overcome some of these limitations. Lewis and Mills (1995) and Sussams (1998) expand on these points and give more detail about good practice in writing notebooks, as does Kneale (1999: 154–9).

The features of a good notebook will be its logical organisation, comprehensiveness, neatness and clarity.

PITFALL 13   ON AVERAGE . . .

'On average' can mean using:

- the arithmetic mean
- the geometric mean
- the harmonic mean
- the median
- the mode . . .

. . . which are all usually different numbers. Check which 'average' is best for the circumstances.

5.6 DISSERTATIONS AND PROJECTS

You can't turn a thing upside down if there's no theory about it being the right way up.
G. K. Chesterton (attributed to)

A dissertation or project may well be a major component in your degree, amounting perhaps to 10 or 20 per cent of the marks on which your degree will be classified; so it is important to know how a dissertation is assessed. The structure of a dissertation and the reasons for asking you to produce one were described in Section 4.7. The criteria for assessing a dissertation or project will reflect these reasons.

1) *Your choice of topic* will be assessed. Credit will be given where it is a novel, feasible and interesting subject, rather than one which has been

covered frequently before and has limited scope to say anything new. A novel approach to an old topic will also be looked on favourably.

2) *The structure of the dissertation* (as set out on its Contents page) will be marked. The dissertation should follow a logical progression, starting from our current understanding of the topic, through your primary (that is, original) research on some aspect of the subject, to general conclusions on this aspect of geography.

3) *The quality of the literature review* which establishes our current understanding of the topic will be checked. Is this review complete and well written using the criteria for essays given in Section 5.4 of this Guide? Does the literature review lead on to the subsequent research and justify why it should be carried out?

4) *Your research methods* will be studied carefully. Have you used the appropriate methods (or more likely the right combination of methods) to research this subject and have you justified your choice of methods convincingly? Have you used the methods correctly?

5) *The analysis of your results* (quantitatively and/or qualitatively) needs to be effective. Have you done this correctly and are your results clearly set out?

6) *In your conclusion*, do you use the results of your research to improve people's understanding of the general area of geography you selected?

7) *Is the dissertation well presented* in terms of its use of English, graphical material and maps? (See Turabian, 1987.)

A very good dissertation is one which scores highly on all these criteria. Flowerdew and Martin (1997) and Parsons and Knight (1995) give fuller treatment of these issues.

Grade descriptors for dissertations (that is, the qualities of a 50%, 60% or 70% dissertation) are given in Appendix F. Your university's descriptions may be a little different from these but they are a good general guide.

5.7 ORAL PRESENTATIONS

In recent years many departments have included oral presentations in their teaching and assessment repertoires. The idea is that students not only need to be able to write well (as tested by essays and examinations) but should also be effective speakers. Much of the cut-and-thrust of working lives consists of discussions, meetings and debates. So you will often find tutors and lecturers requiring you to give talks and assess your performance.

Your department might start in Year 1 by asking you to give a talk as part of the tutorial system – a short talk, in an informal setting, with a small audience and probably not assessed. In later years the presentations may become longer (20–30 minutes), the setting more formal (standing in front of the class), the audience larger (a whole class) and the presentations may be assessed now that you have had some practice in giving talks. The bigger the audience, the more you have to speak loudly enough to be heard; it will become a more impersonal address rather than a talk to a few colleagues.

Practice

Giving a good talk is partly a matter of practice; you will learn from each talk you give and so get better at it. There are a number of ways of quickly achieving some competence in oral presentations. The first is to prepare your material thoroughly – make sure you include all the relevant material that time will allow. In this sense the talk will be assessed much as one would an essay.

Audience

Then you need to consider your audience. Imagine that you are in the audience listening to the talk. You will want the speaker to help you by outlining the structure of the talk right at the start. It will help if they speak clearly, repeat key points while you take notes, and give examples of trends or issues. That can be the model for your talk. How can you get your material across as effectively as possible? Would a handout help for statistical or graphical material, photographs or key facts? Would an overhead transparency (a 'foil') be useful to show the main stages in your talk and perhaps also statistical material? Would slides be useful? Don't use any of these devices just for their own sake; but if you feel they would help you present your ideas clearly, then do use them. You also need to check beforehand exactly how the slide and overhead projectors work. Unless you are a professor, mechanical incompetence does not inspire confidence.

Rehearsal

The next stage is the rehearsal. Run through your talk so that you are fluent with the ideas, words and visual aids. Check the talk is the right length and, if it is not, then add or cut out material. It might be useful to have someone listen to your rehearsal to check whether you are audible and not speaking too fast. The most common faults with talks are that they are delivered too rapidly (and so become an unintelligible gabble) or are inaudible as you mumble nervously to yourself.

After you have carefully prepared your material and practised it, the actual presentation should be a little easier. Try to talk to the audience and not just read your notes all the time. Make eye-contact with your audience occasionally. If something goes wrong or you 'dry up', don't panic – you have your notes, so find where you left off and just carry on without fuss.

If your presentation is going to be assessed – this may be by the lecturer alone or by the lecturer and the class jointly – the likely assessment criteria are these:

* the structure of the talk;
* the completeness and accuracy of the material;
* the presentation itself (audibility, eye contact with the audience, speed of speaking);
* your use of visual aids;
* whether the talk was interesting.

The advice in this section is directly targeted at helping you score highly when your oral presentations are assessed. More detailed advice on giving talks may be found in Appendix G. The articles by Hay (1994) and Young (1998) and Hay's book (1997) are also valuable sources of hints and tips on oral presentations.

The key to public speaking is keeping going even though you may feel nervous.

KEY TIP

✓ **When giving a talk, think of your audience; they (like you) would wish to listen to things which are clear and interesting.**

PITFALL 14 BEING PRECISELY RIGHT

'British' not 'English' (unless you really mean 'England' only).
'The Netherlands' not 'Holland' (which is only a part of that country).
'Businesspeople' not 'businessmen' (unless you mean the men only).
'Farming' (what farmers do) or 'agriculture' (the concept and what governments have policies for)?
'Rainfall' or 'precipitation'?
'Harbour' (open to the sea/lake and with variable water level) or 'dock' (an enclosed water area with stable water level).
'State' (a high-level political entity) or 'nation' (which may not be politically separate although the people may feel themselves separate or different).

Use the precise word for the sake of accuracy and to avoid giving offence.

5.8 POSTERS, WEB POSTERS AND PRESS REPORTS

Posters

A poster is one of the less commonly used assessment methods, but you may be asked to produce one, perhaps as a member of a group. The idea is to convey some geographical material to a lay audience, as opposed to essays and dissertations which are specially written in an academic style for an academic audience. So a poster will comprise short blocks of text, photographs and graphic material with a very simple linear structure which can be picked up and followed through quickly by readers who are virtually passers-by. They will be looking at your poster on a wall from a distance of two or three feet. You need something eye-catching but not off-putting. So the poster has to be bold and clear. It should be attractive to look at and structured so that the reader's eye is led through it from start to finish. You do not want a lot of text or a complex structure. You may try something along these lines:

- here is an issue;
- here is how it manifests itself and why it is important;
- here is what can/should be done about it.

Not all geographical subjects lend themselves to the poster format. Suitable ones are those which are current public issues, such as a proposed new road, global warming or the incidence of crime. Hay (1997) and Vujakovic (1995) give more examples of good poster writing.

A development of the paper poster is a web poster. The principles are the same – being concise and clear, and using visual material effectively. Additionally, you can add hyperlinks to other websites so a web poster is potentially 'expandable' and, with the right computers, can include sound tracks and moving images as well as static text and pictures.

"You don't think you might have taken the idea of the interactive poster a little too far?"

P. Vujakovic '95

Press reports

A different but related task is to write a press report on an issue. You might be asked to write an article of, say, 1,000 words in the style of a quality newspaper. Again, the aim is to teach you how to write in a non-academic style without losing accuracy or intellectual integrity. The journalist's style of writing is distinctive so, to do a good job, you first need to read newspaper articles unusually closely, to analyse the use they make of, for example, short sentences and paragraphs. The first paragraph of a newspaper report sets out the issue immediately and clearly (it is not an introductory preamble as you would use in an essay). A press report might include quotations from those affected by the issue. To do well with a press report exercise you need to absorb quickly a new style of writing and yet still produce a fair account.

5.9 PEER ASSESSMENT AND SELF-ASSESSMENT

Sometimes departments will ask you to mark another student's work (peer assessment) or to mark your own work (self-assessment). This may seem strange at first – it is the staff's job to do the marking, isn't it? However, there is a point to this. To be able to mark someone else's work fairly you have to know both the subject and what the features of a good performance in that kind of assignment are. The important point is to be able to justify your comments and your mark. Once you have identified what makes other people's academic work good, you can use these insights to improve your own work. In later life, being able to judge things fairly is an important skill. At first it is unnerving being a marker, but you will get better and more confident at it.

Self-assessment is similar. It is really a learning experience and a chance to reflect on where you feel you have made good progress and where you got stuck. Comparing your assessment of your work with your tutor's is instructive. Students are usually surprised at how close the agreement is. Even better, try assessing your work *before* you hand it in; you may be able to improve it.

5.10 HOW TO REVISE

You need to revise for your examinations if you are going to do well, but how should you revise? Is there a best way of revising?

Perhaps the starting point is how you currently do your revision. If that pattern works well for you, then keep using it. If you are doing less well in examinations than you wish (and less well than you are doing in other forms of assessment), then you may want to think about other ways of revising.

A timetable for revision is always a good idea. Work backwards from the dates of the examinations themselves and plan blocks of revision for each one. The amount of time for each examination depends on how many there are and when you are free from teaching and coursework to concentrate on the revision. You will have to calculate that yourself each year.

Examination answers are short – only 700–1,000 words for a one-hour answer and less for 45-minute answers. An essay might be three or five times as long. You have time in an examination to write only a limited amount, so it had better be the key points. Facts, details, examples, quotations or statistics are useful devices to elaborate general statements, but the key point in many examination answers is for you to show that you understand the main issues and arguments as put forward in the literature. So revising the main points right across your course – getting the big picture – is better than revising just a few areas in great detail. It clearly saves time if you have done most of the important reading well before the revision period starts and so you can then focus on reviewing the notes you took on that reading. At the end of a block of revision, ask yourself what are the five key points you have learned on this topic, and jot down the points – in effect, making notes from your revision notes. Actually making notes helps embed the material in your mind.

One of the main problems with many examination answers is that they lack a good structure. The student displays knowledge of the topic but it is in the wrong form or only partially relevant. To help overcome this mistake, you can try practising the first five minutes of examination answers. In the first five minutes you will be analysing the question, deciding what you know about it and planning the structure of your answer, perhaps in pencil, as a guide for your actual answer. Those first five minutes can be crucial because a good structure gives you a platform for a good answer, while the wrong structure cannot be salvaged no matter how much you know about the subject. Most university libraries and departments have a collection of previous years' examination papers which you can use as a source of questions to try out (provided the syllabus has not changed, of course). Practising the planning phase is really useful. Actually writing out the full answer in the remaining 40–55 minutes should be straightforward if you stick to the plan.

Some people find that revising with another student taking the same examinations is helpful. You can test each other on what you have learned, section by section, and answer each other's questions. Saying the

THE USE OF BUILDINGS AND AREAS CAN BE REVISED DRAMATICALLY. REVISIONS CHANGE THINGS.

FIGURE 12 *Musée d'Orsay, Paris, France – urban railway station into international art gallery*

FIGURE 13 **Dodge City, Kansas, USA – wild-west cattle town into tourist centre**

ideas out loud helps you remember them better than just silently reading them. Revision can be a very lonely task, so company may help; a lot depends on how well you get on with your colleague.

Revision is also a physical and mental strain, so getting enough sleep, some exercise and punctuating the long revision hours with short breaks and nice treats is essential to keeping you going. You will come to revision fresher and in better spirit, and that is more important than endless hours of unproductive revision, learning little and remembering less.

5.11 HOW TO SURVIVE AN ORAL EXAMINATION

An oral examination (no, it's not dentistry but it can still be an uncomfortable experience) means an examination where you are asked academic questions in person by an examiner. It is sometimes called a *viva voce* examination or a 'viva' for short. Vivas are not common but you might be called to one. They are held for various reasons:

- to check that you actually wrote a piece of work if there are doubts about this;

- to deal with any disputes over the quality of your work;
- to fill in any gaps in your record of assessed work;
- to decide on which class of degree you should get if your final marks are right on a borderline.

The last of these is the most common and is still found in the older universities for final-year students.

The set-up is this. Right at the end of your final year, you may be asked to attend an oral examination with the external examiner for your degree scheme. He/she will be a senior geographer from another university who has been employed by your university to ensure the maintenance of academic standards in your department. You will probably be told whether you are having a viva only a day or two before it takes place, so you need to be around the department then. You can ask who the external examiner is and what his/her specialisms are. You can also ask which borderline you are on (is it the lower second-class degree and upper second borderline, or the upper second/first border?). Usually an oral examination is used for people just below a class boundary. So if the oral goes badly, you stay in the lower class of degree; if it goes well, you move up a class. Hence there is everything to be gained and nothing to be lost from a viva.

It is hard to prepare for a viva because you cannot anticipate what you will be asked during the 15–30 minute examination period. Often the external examiner will get you to talk about your dissertation, so you could skim over that the evening before the viva. A general essay paper might also be the subject of questioning, or indeed any of your other essays, so you could try to remember which questions you answered and roughly what you said. Finally issues in the news might be the focus for questions, so think what the geographical perspectives are on current affairs.

During the viva itself, the key points are these:

- accept that everyone is going to be nervous in their viva and then control your nerves as best you can;
- pause and think before you answer a question;
- try to be as fluent as you can.

Rightly or wrongly external examiners tend to be impressed by students who:

- are honest enough to admit when they don't know something;

- can go beyond what they wrote in essays, dissertations, etc.;
- can criticise their own work;
- can make connections between courses;
- show a knowledge of the geographical literature;
- link geography to current affairs.

After the viva you will probably not know how it went; the result will be your final class of degree.

If the viva is being used at an intermediate point during your degree to check on your work (although this is unusual), you have the right to ask for full details of which elements of your work are being re-assessed and why an oral examination is being used for this.

PITFALL 15 EXCEPTIONS

'This is the exception which proves the rule.' Of course, logically, exceptions cannot prove (in the sense of 'confirm beyond doubt') any rule. The word 'prove' in the maxim is used in its older sense of 'test', as in 'to prove oneself in combat or a viva'. Exceptions test rules to see how well they stand up to contradictory evidence. For Karl Popper, research was about collecting as much contradictory evidence as one could and seeing whether there was enough to falsify our current understanding of how the world works. You can try to falsify statements; you can never prove any statement to be absolutely true.

5.12 MARKS, CREDIT AND DEGREE CLASSIFICATION

Marks

For each examination and piece of coursework you will get a mark. What usually happens then is that the marks from the examination, essays and other coursework for each course/module will be amalgamated to give you an overall course/module mark. Your course/module guide should tell you the relative weighting given to each piece of assessed work (e.g. 50:50 examination and essay; or 70:30 examination and practical exercises).

Credit

Traditionally a degree scheme (and individual courses/modules within it) were defined in terms of their length – how long it would take to complete the degree scheme or course. So, traditionally an English or Welsh first degree (BA or BSc) is three years long, and an individual course or module might run for ten weeks, two terms or one semester. This is an unhelpfully rigid system because it requires all students to complete all their work at the same 'speed' as everyone else. For a three-year degree you have to start at the same point as everyone else and the degree has to be completed in three years, no quicker and no slower. Only material taught within that degree scheme can count towards the final award – prior learning is not valid. The system of credit weighting tries to overcome these rigidities.

Each course/module is given a credit weighting, depending on how long it is. To be considered for the award of a given degree you have successfully to complete sufficient courses/modules to achieve a given total number of credit points – usually 120 credit points per year for a BA or BSc, 360 for the whole degree. Each university will have its own regulations about how many of these credit points have to be on certain topics, in specified subjects and at final-year level.

The advantage of a credit system is that it allows the degree to be built up to the credit points target over a variable period of time. So you can take more than three years to complete a BA or BSc if your family, work or financial circumstances require a more extended period of study. Equally, credit may be given for courses studied at another university or even for other types of education at an appropriate level. This allows you to transfer between universities and still get a degree. The rules governing credit are usually complex and vary among universities, but they may be worth your while investigating locally if this pattern would suit you.

Degree classification

Finally your university will bring together all your course/module marks to guide the Board of Examiners towards your final class of degree or grade point average. Each university does this differently so you will need to find out how it is done in your institution. There will also be different rules on what happens if you fail a unit, need to re-sit assessments or wish to appeal against your degree result. Again, it is worthwhile to find out about these local rules and regulations – hoping, of course, that you never need to use them!

So far we have looked at how to make the most of the formal parts of your time at university – learning geography and being assessed in it. There is obviously a great deal more to life at university than formal study, and in the next chapter we look at how to benefit from the rest of your time there.

6 OTHER USEFUL ACTIVITIES FOR GETTING A JOB

A geography degree is (or should be) hard work, but it certainly should not take up all your time at university. A fair balance of study, sport, paid work and leisure will be beneficial and much more fun. There should also be time outside the formal structure of your degree to broaden your horizons and perhaps bolster your chances of getting a good job. There are various ways you can do this:

- moulding your degree and choosing courses;
- acquiring new skills;
- jobs;
- active citizenship;
- networking and contacts;
- leadership and teamwork;
- using your careers service.

6.1 MOULDING YOUR DEGREE AND CHOOSING COURSES

Departments of geography usually offer several degree schemes. A 'Geography' degree will normally train you in all the major aspects of the subject. In Years 1 and 2 the extent to which you can choose your own courses may be rather limited. You will probably have to study a core of compulsory courses which will give you a broad foundation across geography in terms of subject knowledge and skills and will lay the foundations for more specialised courses. In the later years of your degree you will have more opportunities to select courses so that your degree scheme can be weighted to particular aspects of the subject – for example, physical geography, environmental topics or quantitative methods – or towards particular forms of teaching that you like (such as fieldwork or computer applications).

More radical choices can be made if your university offers separate degrees in Human Geography, Physical Geography and Environmental Management. In these degree schemes you can focus more fully on one section of geography, perhaps cutting out other aspects which appeal less to you.

In many departments you can also study for a joint degree or a combined honours degree, half of this being in geography and the other half in another subject – a foreign language, perhaps, or geology or economics. A joint degree gives you less geography, but you will have *two* sets of knowledge and skills as opposed to the geography student's one. That gives you a breadth of abilities which many employers might find very attractive, as well as giving you greater variety in your higher education.

Universities vary quite a lot in the range of geography degrees they offer, and in the extent to which you can choose your courses to mould the degree to your needs. This is worth checking before you finally choose your university. Universities also vary in their willingness to allow you to alter your degree after you have started it. In some universities, no matter how much you dislike your initial choice, you have to stick with it throughout your time at university. In others, the ability to change your degree scheme is greater. This too is worth checking before you finally choose your university.

NATURAL RESOURCES ARE ALWAYS BEING RE-ASSESSED AS TO THEIR SOCIAL AND ECONOMIC VALUE.

FIGURE 14 *Abandoned house, Dun Carloway, Lewis, Scotland*

FIGURE 15 *Coastal defence barrage, south-west Netherlands*

6.2 ACQUIRING NEW SKILLS

Having got this far through the Guide, you will appreciate that a geography degree can fairly claim to be teaching you a wide range of skills and that this is a clear strength of geography as a subject. It is also an interdisciplinary subject and will give you a mental flexibility denied to many other graduates. However, there are other skills which you will not be taught. Computer skills are so broad and fast-moving an area – and one which is increasingly important in so many sectors – that the best advice must be to acquire as many computer skills as you can. You may learn about other software packages at home, from your parent's workplace, summer vacation courses, short courses put on by your university, or during vacation jobs. However you learn them, some familiarity with a wide range of programs and packages will be a useful addition to your curriculum vitae or résumé. You never know when they may come in useful.

Another set of skills relates to foreign languages. The global economy and the integration of Europe mean that foreign-language skills, even at a basic level, will be increasingly appreciated by employers. You can learn foreign languages from the university language laboratory, self-instructional packs, evening classes or when abroad. This is a useful addition to your profile, provided that you do not claim on your curriculum vitae or résumé more language skills than you actually have.

Other skills that might prove valuable would be driving, first aid, and certificates in leadership, supervisory skills or personnel management. Some of these might be developed through schemes such as (in the UK) the Duke of Edinburgh Awards or working with youth or community groups.

It is useful and fun to be able to turn your hand to many things. For fast-moving careers in the future, the key skill may well be the ability to cope quickly with new situations.

6.3 JOBS

You will probably take vacation or term-time jobs to earn money. It is worthwhile thinking about how you can use these jobs to improve your career prospects. At the simplest level you can expand on the 'previous employment' section in your curriculum vitae or résumé which, if you are only in your early 20s when you graduate, is inevitably going to be rather short. You can capitalise on a *variety* of jobs to demonstrate that you understand what different employers are looking for in their employees. Most jobs, even fairly menial ones, require punctuality, attention to detail, honesty and being a responsible employee with a good (that is, low) absence record. You might have had some responsibility for cash-handling in this job, for dealing with the public or supervising staff more junior than you. Analyse your jobs and see what you learned from each of them. Of course, if you have had full-time jobs before coming to university, you can include a great deal more under 'previous employment'.

Aside from traditional student jobs you could try for ones which will give you skills directly relevant to employers – working in sales, stock control, computer-based jobs, or an administrative post. These skills will immediately appeal to employers. You could try targeting jobs in the areas where you want to gain employment after graduation. So, if your career goal is environmental work or teaching, for example, then jobs in conservation or working with young people would establish your commitment to the area and give you an appreciation of the realities of the sector's basic activities. If you have a specific company in mind, try to get a summer job with them. Even if it is at a rather junior level, you will still gain first-hand experience of the firm and they of you.

ACTIVITY 16

Note here and in your Personal Record (Appendix B) what you have learned from each vacation or other job you have undertaken.

Job 1 From this job I learned . . .

Job 2 From this job I learned . . .

Job 3 From this job I learned . . .

6.4 ACTIVE CITIZENSHIP

Some employers are impressed (and so they should be) by people who, besides studying for a degree, have also given something back to society through voluntary or charity work. This can take many forms – fund-raising, being a counsellor on a student-run Nightline or advisory service, practical conservation work, helping at a shelter for the homeless, running a Sunday school or cub/brownie group, or teaching English to non-native speakers. These are all intrinsically useful and valuable activities. You at least will gain a lot from it. You will get an insight into other people's lives, a sense of making a contribution to society or a good cause, and a chance to do something completely different from your university studies. You will also meet a new circle of friends and contacts. Why not tell prospective employers about your voluntary work?

6.5 NETWORKING AND CONTACTS

This is probably the most intangible of the things you can do to improve your employability. It does not involve any single action common to all students. Rather, it is a suggestion that 'who you know' may be as important as 'what you know'. Nowadays there are usually formalised job-application procedures which large appointing committees can use to evaluate objectively competing job applicants, and this means that in large sectors of the economy personal contacts alone will not guarantee you a job. However, knowing people already working in your intended career area may benefit you in other ways. You may get to hear of opportunities sooner or pick up ideas on how best to present yourself so as to meet an employer's current needs. Your contacts could give you guidance on what working in the sector or for a specific employer is really like – something the career guides cannot tell you.

The tricky part is gaining the contacts. Friends and family may help. Your vacation jobs may put you in touch with people who can advise you. Your geography course may have elements of work-based learning through which you will meet useful people. Similarly, you may choose your dissertation topic so that you come into contact with people in your proposed career area. Your university's Careers Service or Department may have lists of previous graduates of your university (its 'alumni') who are available to give advice. Your university's alumni association may have a mentoring scheme which puts you in touch with former graduates or it may run career fairs on campus at which you can meet such people. You can look in specialist journals in your field and contact the authors of articles where a contact address is given. Introduce yourself and ask about work opportunities.

ACTIVITY 17

Have you started building up your own network of contacts through your Department's staff, careers service, potential employers or through your family, social contacts or work experience?

Who are these people?

What might you **and they** gain from such contacts?

6.6 LEADERSHIP AND TEAMWORK

It is arguable that as organisations 'de-layer' and become 'flatter', more responsibility than previously is being placed on those at the lower levels of the management pyramid. It is being given to those at ever earlier stages in their careers – that is, to people like you within five years of graduating. We explored these ideas in Sections 3.2 and 3.3. So employers are looking to recruit people who have already demonstrated some signs of leadership or leadership potential and who can work in successful teams.

So, what is 'leadership'? It includes:

* motivating a team of people to work to a plan enthusiastically;
* understanding that successful groups involve everybody, they don't just take orders;
* supporting all the team members;
* organising people, events, resources and timetables;
* securing resources;
* being accountable and responsible.

How might you convince a sceptical recruiter that you are a leader and a good team member? On the basis that 'actions speak louder than words', it would be useful if your curriculum vitae or résumé records examples of you as a leader or organiser, either by yourself or with a group of other students. This might include your role in a college or student society or an organisation that is a part of your home life, such as a sports club, religious organisation or social society. Whenever you claim leadership or organisational skills, you need to back this up with specific examples of what you have done. Employers also want team-players so it is just as important to have helped to run something as to have done it all yourself. Being good at teamwork is a respected skill.

KEY TIP

✓ **Help organise things, particularly in areas that mean a lot to you personally – sport, hobby, church or politics, for example.**

ACTIVITY 18

What activity or event have you organised or helped to organise in the last six months (e.g. a sports event, social occasion, team project, club or society activity)?

Are there any lessons you have learned from organising such events that would help with future events you might run?

Note in your Personal Record (Appendix B) what you organised and how well it worked.

6.7 USING YOUR CAREERS SERVICE

Your university's careers service is an invaluable source of advice and training on careers. Here is a list of some of the things it can do for you:

- its library will tell you about types of job and specific employers;
- it can train you in specific job-search skills, such as how to write a curriculum vitae or résumé and how to do well in interviews or aptitude tests;
- it can provide one-to-one counselling on your career choices;
- it arranges careers fairs and interviews with potential employers;
- it provides computer programs that may help you decide on the kinds of job you wish to go for.

Your careers service is also a gateway to various online sources of information and services whose printed equivalents will be found in the careers service's library. For example, AGCAS (the association of careers services in the UK) publishes a lot of useful material on career issues at http://www.agcas.csu.ac.uk/publicati/menu.htm.

Details about individual occupations, including teaching, can be found at: http://www.prospects.csu.ac.uk/student/cidd/occupations/main.htm.

Links to the websites maintained by various professional bodies (so you can check on their advice to potential entrants) is available from: http://careers-main.lancs.ac.uk/org-links.html.

If you want to check which careers may suit your career aims and personal values, you can try 'Quick Match' which matches the features you want from work with what different careers can offer. The website is at: http://prospects.csu.ac.uk/student/cidd/main.stm and click on the Quick Match button.

This website (http://prospects.csu.ac.uk/student/cidd/main.stm) also provides details on alternatives to UK employment (such as work overseas and self-employment), and guidance on actually applying for jobs (e.g. the application process and being interviewed), and there is a set of review mechanisms for your skills and aspirations.

There is advice on job hunting using the Internet at: http://careers-main.lancs.ac.uk/findinet.htm which includes search engines and a range of information sources.

In Sections 6.9 and 6.10 we give some advice for those who see postgraduate study as their next step after graduation.

Many students make too little use of the many facilities that their university's careers service can offer or do not investigate it early enough. Certainly do not leave it until your final year at university before exploring how it can help you.

Also useful are the various books on how to go about finding a job which you can buy in any good bookshop, such as the one by Bates and Bloch (1997).

6.8 A GEOGRAPHER'S CURRICULUM VITAE

A curriculum vitae (a CV) or résumé is a statement of your achievements so far (and not just your academic ones) that you send to potential employers along with a completed application form and/or letter of application. A geographer's curriculum vitae will look rather different from those of students in other disciplines simply because of the different balance of courses and teaching methods typically found in geography departments. In this section we set out some of the elements you can bring into your own CV.

Key points

To be effective, a curriculum vitae needs to be truthful, well presented and as full as possible. The most common fault with CV's is that they sell the applicant short because the person has not realised quite how many skills and achievements he/she has already acquired.

So, the starting point for writing a CV is a basic list of all your activities, skills and experiences, such as the 'Personal Record' set out in Appendix B. Remember that this list includes not only your academic skills and achievements but also those you acquired before you came to university, those developed by any jobs you have had (even modest summer ones) and the things you have learned from the more social side of your life, such as sport or the clubs you belong to. Keeping your Personal Record updated (15 minutes a term should be enough) is obviously helpful when you come to write your CV.

So, in this chapter we shall provide:

1) A specimen CV to show you the sorts of things to include and how to arrange and display them;
2) A reminder of the sorts of skills you are very likely to have acquired while pursuing a degree in geography.

A template for a curriculum vitae

John Smith B.Sc.

Personal details

Addresses – home, university, work
Telephone numbers, fax numbers, e-mail addresses
Date of birth
Nationality

Educational history

School(s)   dates, names, places
            school examination results (at A level or equivalent only)

University   first degree   dates, name of university, degree title, result
                             (if known)
                             note specific options or parts of degree of
                             relevance to each application
                             note any 'outside' subjects and special
                             strengths within the overall results.

Work experience

Give major relevant jobs with dates – do not leave 'gaps'. Give more details on full-time jobs as opposed to holiday jobs. Briefly describe the job if this is not obvious, e.g. 'Administrative Assistant – in charge of visitor education and publicity'. Show what you gained from each job – 'in charge of cash handling' or 'dealing with awkward customers'.

Personal skills

List here the personal qualities and skills which your education and work experience have given you (for example, supervising other people, time management, working to deadlines, producing good-quality work, dealing with the public, understanding how industry works, giving presentations, etc.). Stress those skills that are particularly relevant for each job application. Remember to show how you got the skill (e.g. problem-solving from doing my dissertation, working in a team from doing the group audit, dealing with the public from your barperson job).

Computing skills

If you have a range of computing skills (such as word-processing packages, Excel for spreadsheets, a statistical package, a database package and a graphics/GIS package), then make a feature of your computing skills here. Otherwise, list them in the 'Other relevant skills' section (below).

Languages

If you are fluent in some languages, then make a feature of your language skills in a separate section. Otherwise list them in the 'Other relevant skills' section (below). Even a GCSE course can be worth recording.

Other relevant skills

Include here any other skills that are relevant for the job application, e.g. first aid, sports training, driving, foreign languages and computing skills if not recorded elsewhere.

Personal interests

This is the human-interest part of the CV. Avoid bland things like 'reading' or 'travelling' or over-the-top claims like 'saving the planet'. Unusual things like parachuting or Thai cooking may be interesting, but avoid what might be interpreted by some narrow-minded recruiters as odd (Satanism or breeding ferrets) or worrying (going clubbing). If you have travelled a lot, give details here of where you went and what you gained from the experience. Work you have done on the Duke of Edinburgh Award scheme or for charities could be given prominence in this section.

Referees

Line up three referees, at least one an academic geography referee, perhaps your dissertation supervisor or tutor. A 'work referee' is also very useful. For each, give his/her full name, title, position and contact address (postal address and telephone number at least). You will probably not need more than two referees for any application but you may need different types of referee for different applications. Give your referees a copy of your CV so they know what you are saying about yourself.

KEY TIPS

✓ **Keep the CV to three pages – two, if possible.**
✓ **Avoid spelling mistakes.**
✓ **Make your CV distinctive by designing your own rather than using a standard template, as is found on some word-processing packages.**
✓ **Lay out the CV neatly but do not overdo the formatting.**
✓ **Avoid humour.**
✓ **A good CV tells a story about you and is interesting to read, rather than just a list of facts about yourself.**

What will distinguish geographers' CVs from those of other students will be the wide range of specific skills that a geography degree will allow you to claim and demonstrate. The next section suggests what those distinctive skills are.

What you get from a geography degree

Geography is a distinctive subject in that it draws upon and provides a link between the social and physical sciences with their particular approaches and skills. Geography allows you to train in both human and physical geography, to explore the interrelationships between the social and physical sciences, to develop skills in research design, statistical analysis and computing and, if desired, to specialise in human or physical aspects of the course.

Successful completion of a geography degree should help you develop the skills in demand by many employers of graduates in managerial, professional and scientific work. These skills include:

- the ability to work independently or as part of a team;
- the initiative to formulate and investigate a problem using field research or secondary sources;
- facility with verbal, numerical and graphic modes of communication;
- familiarity with modern information technology (such as computers);
- a good background knowledge of social, economic, environmental and political factors in Britain and elsewhere in the world.

Some employers, for example in marketing or transport, may look specifically for geographers; in other jobs, such as planning, teaching, cartography, environmental management and conservation, a postgraduate course following a geography degree offers the usual route of entry.

The majority of geography graduates, however, go into jobs for which a geography background is not specifically required, but for which the skills gained in a geography degree are a positive advantage. These jobs include accountancy, banking, computing, insurance, local government, civil service, research, the armed forces, the police and various types of management. Geography graduates fare at least as well as graduates in other disciplines in the present job market and a geography degree can lead to a much wider range of career options than some more obviously vocational degree courses.

As a geographer you have a great deal to offer employers. The following checklist reminds you of the range of skills, experience and knowledge which you have and which you can draw to the attention of potential employers in your application forms and CV. Not all the topics will be equally applicable to everyone and every job, but aspects of this checklist should be helpful for most people.

Literacy
Library skills to collect information; writing clear prose for essays, project reports, briefings and the dissertation; ability to think critically and argue logically.

Oral skills
Fluency in oral discussion (as in seminars and presenting papers).

Numeracy
Statistical training; analysing data from small field surveys (such as your dissertation) up to the manipulation of very large datasets (e.g. the Population Census).

Computer skills
Word-processing; use of PCs; file handling; keyboard skills. Mention spreadsheets and use of the web.

Graphical skills
Ability to draw maps, graphs and diagrams; handling spatial and areal data as well as maps and aerial photographs; GIS and remote sensing skills.

Time management and organisation
Working to deadlines; working with teams of students on projects; liaising with officials and the public while on projects; orderliness; attention to detail within a big programme of work (like the dissertation); ability to work quickly and under pressure.

Working independently
Your dissertation is a solo project of considerable size and complexity; the need to be self-reliant and self-motivated; the ability to organise work, solve problems and avoid (or cope with) crises.

Working in a team
Your experience of team-work will include projects and fieldwork, and it may be more extensive than that of students in some other departments. In teams you learn tolerance and the ability to persuade others to work to a plan.

Technical skills
In practicals, on field courses and in some options you will acquire a range of specialist skills in data collection, analysis and presentation using a variety of equipment and methods (for example, surveying, till-fabric analysis, household surveys).

Knowledge of the world
You will acquire a knowledge of particular parts of the world and of contemporary social, economic and environmental issues – employers appreciate such awareness.

Breadth of degree
Geography is a broad subject and your degree scheme will reflect this in its combination of theoretical and empirical material, human and physical courses, analytical and practical skills.

Other skills
What other skills have you got – training, leadership, driving, languages, etc.?

Other activities
All employers like to see applicants who have taken part in activities outside the degree scheme – hobbies, sports, societies, expeditions, voluntary work, work experience in the vacations. What can you offer here? What have you learned from these activities?

6.9 FINDING OUT ABOUT POSTGRADUATE COURSES AND RESEARCH OPPORTUNITIES

You will have seen from Section 3.4 that many geography graduates go on to further study after their first degree. This will be a focused vocational course which provides specific skills for a clearly defined job. In some cases a first degree (in geography, for example) will be needed to gain access to this further training. In other cases the course might be put on by a university or a professional body which maintains the standards of those entering the profession.

Taught postgraduate courses

These courses lead on to Diplomas and Masters degrees. They are offered by many universities and are highly relevant (even essential) for careers in planning and environmental or conservation work, for example. To find out about the entry requirements for various career areas you might look at: http://careers-main.lancs.ac.uk/org-links.html which provides links to the websites of most professional bodies in the UK. This will tell you whether further training courses are needed for a profession and, if so, which ones.

If you want to know which postgraduate courses are available you can look at: http://www.prospects.csu.ac.uk/pg/ which is a database you can search by university, subject or keyword.

Of course you need also to fund yourself through this period of further study and training and the same website (http://www.prospects.csu.ac.uk/pg/) gives access to a database of opportunities to gain funding for postgraduate study. Also useful is: http://www.lifelonglearing.co.uk/cdl/ which describes career development loans.

If you want to find distance learning or open learning courses, a good website to try is: http://www.distance-learning.hobsons.com

More details on MBA courses are available at: http://www.mba.hobsons/com and on teaching from: http://www.prospects.csu.ac.uk/student/cidd/ by entering 'teaching' in the search box.

If you want to study outside the UK, then this website: http://www.studyoverseas.com is a good starting point, while for courses in the USA try: http://www.petersons.com.

A way of accessing the websites of all UK higher education institutions is the screen-sensitive map of the UK at: http://www.scit.wlv.ac.uk/ukinfo/ which will give you more details on particular universities, departments and courses.

Research opportunities

To find out about opportunities for research leading to MPhil or PhD degrees you can use many of the sources listed above for taught postgraduate degrees as well as advice from your department.

Details of research opportunities in the UK and sources of funding are available at: http://www.prospects.csu.ac.uk/pg/

Links to the universities' websites are available at: http://www.scit.wlv.ac.uk/ukinfo/

6.10 HOW TO WRITE A RESEARCH PROPOSAL

If you want to study for a postgraduate research degree (a PhD or M.Phil) then you are going to have to write a research proposal which sets out what you want to do research on. Your potential supervisor and funder will need this information. What should a research proposal look like? What form should it take?

The structure suggested below is a general one, which will be a useful starting point for many students. You will need to 'customise' it to reflect your specific topic, progress and circumstances. The length of each section will vary, and the lengths given in brackets after each section are just for initial guidance. Your potential supervisor – do talk to staff about your plans as they will be able to help you a lot – may suggest a different structure or changes of emphasis

Introduction

A brief setting of the scene to your research area; its current policy importance; its wider theoretical importance; the structure of the rest of the proposal (1 side of A4 paper).

Literature review

For your research area, what are the key ideas and developments; the policy literature; areas where you see weaknesses or new areas needing research; identify the relevant 'research frontier' from which you will start; how your research will be an original contribution to knowledge and understanding (2–3 sides of A4).

Your research

Stage by stage, describe what you will do; roughly how you will achieve this; describe possible methods, noting those which will probably be feasible and those whose feasibility needs further checking (2–3 sides of A4).

Analysis

How might you analyse your findings? (½ side of A4).

Research findings

Roughly what form might your research findings take? (½ side of A4).

Timetable

Give a draft timetable which shows how you will achieve all your aims within the period of your research (usually 36 months for a full-time PhD).

Issues

Note here any issues concerned with safety, confidentiality, data protection, intellectual property or ethics.

Support

Note here any issues concerned with resources for your research, your training or procedures.

Different funders and universities will require the proposal to be set out in different formats and lengths, and this general layout can be moulded to their requirements.

If you want to read further on this, we suggest you try Phillips and Pugh (1994: Chapter 7).

At times your geography degree may seem to dominate your life at university, but it will provide only a modest part of the profile that you will present to potential employers. You will need to take the rest of your life and personality just as seriously, not least because they are what make you a whole person and they will add so much fun to your time at university.

WHERE HAVE WE GOT TO?

In this Guide we have tried to show you:

- how you are likely to benefit from going to university (Chapter 2);
- how changes in national educational policy and the needs of employers are combining to encourage changes in university teaching, and the kinds of career geographers have followed recently (Chapter 3);
- how departments will teach you geography, why they use these methods and how you can gain the most benefit from these forms of teaching (Chapter 4);
- how departments will assess you – with some advice on how to do as well as you can under these various forms of assessment (Chapter 5);
- and the ways you can improve your employability and life skills outside the structure of your geography degree scheme (Chapter 6).

This chapter summarises the Guide.

LEARNING GEOGRAPHY

General

- Learning geography is not about prodigious feats of memory. Think of 'facts' as being like the lights on a Christmas tree; there must first be a structure of ideas which can support, and be illuminated by, the 'facts'.

- There are few absolute 'facts' in geography; most geographical knowledge is debatable, provisional and liable to change or be changed.
- 'How you learn' is an interesting topic; Section 2.12 helps you analyse this and improve it if need be.
- *'Reading for a degree'* – two or three hours of reading are needed for each hour of lectures.
- There are different styles of writing – essays, field notebooks, posters, dissertations. Make sure you use the right style for each occasion.
- Have you changed the way you study since starting at university? If so, how?
- *Ask* when you don't understand something; you certainly won't be the only one wanting to know.
- Mould your degree to your career aspirations.
- Do the courses you *enjoy*; you will probably learn far better.
- Don't compartmentalise your geography; think *across* modules and topics.
- Keep up to date with current affairs and consider their geographical implications.
- Your fellow students are people to work with, not compete against.
- You will come to see things differently at university – different questions and different perspectives.
- Expect geography at university to be very different from the subject at school – now it is 'learning' not 'teaching', and being critical of arguments and not just accepting them.
- Keep an eye out for the subject's seven key themes: scale; spaces and places; systems; spatial variation; environments and landscapes; change; and differences.
- Work as hard as you need to.
- If you have a complaint, make it through the designated channels.
- You and the staff have a joint responsibility to ensure your safety.

Lectures

- The most important thing to do in a lecture is *not* to take notes. It is to identify what is really important in the lecturer's words and then record these points – *active listening and summarising*.
- Lecture *notes* – not a transcript.

Tutorials

* Tutorials can be useful *if you prepare for them and participate*. You will get nothing from them if you don't join in.
* In tutorials you can be critical of established authors.
* In tutorials you can practise working and debating with others – a useful life-skill.

Oral presentations

* Oral presentations can be nerve-racking; the key is to keep going, speaking clearly and not too fast. Speaking in public does get easier the more of it you do.
* Practise your talks beforehand with a friend as your 'audience'.
* Keep it simple and clear.
* Talk to your audience not to your notes.
* Think of the best lecturers you have heard and copy some of their style.

Fieldwork

* Fieldwork is fun and also a chance to learn new skills.
* Fieldwork prepared by your tutors provides a good role model for your own research for projects or your dissertation.
* Working successfully in the field with a group of students can be tricky, but it is a skill you will need in later life.

Practicals

* Practical classes can be fun and imaginative.

Library and ICT

* Find out as early as you can how your library and computing systems work; exploiting these to the full will help you greatly.
* Use the Internet's facilities as much as you can, but be as sceptical of what you find on the web as you are of your other reading.

BEING EXAMINED IN GEOGRAPHY

Examinations

- Examinations test your *breadth of understanding* of a large section of geography.
- Most poor examination answers are due to one of several common faults (see Section 5.3). Try to avoid these.
- Practise the planning of examination answers.

Essays

- One of the key points with essays is to understand the command word; check the explanation of essay command words in Section 5.4.
- The essay references you need are missing from the library. Should you:

  o give up and go home?
  o write the essay without any reading?
  o browse the library shelves, catalogue and the web for other material on the same topic?

- Remember always to acknowledge the source of any material you incorporate in your essays and dissertations, and so avoid the charge of plagiarism.
- Make sure you get enough specific and practical feedback from your tutors on your assessed work. If you need more, ask for it.

Your dissertation

- A dissertation will need planning, thought, advice and stamina.
- Here are eight general benefits from a successfully completed dissertation which you can include on your Personal Record:

  o time management skills (you got it in on time);
  o determination not to be beaten (you did it!);
  o negotiating for access to data, land, equipment or interviewees;
  o planning a complex piece of work;
  o solving problems;
  o independence (you saw it through yourself);

- o the new skills you learned for this dissertation;
- o the presentational skills for a major report.

- Have you thought how your dissertation could put you in touch with potential employers and establish your credentials as knowledgeable in your career area?
- During your dissertation, you are in charge; make the most of the opportunity.
- A dissertation is an excellent place to show how good an all-round geographer you really are. What are you planning to do that will impress the examiners favourably?

Revision

- Timetabling your revision is obviously important.
- This is when *concise* notes from your lectures and reading come into their own.
- Practising the first five minutes of an examination answer (i.e. the planning stage) is a good idea.

Oral examination (viva voce)

- Keep icy calm; focus on each question; think before you speak; be critical of authors and yourself.

EXTRA-CURRICULAR ACTIVITIES

- Keep a fair balance in your life – studies, paid work, leisure, sport, family and friends.
- Collect as many basic computer skills as you can, even if some are not immediately relevant. You never know when they may come in useful.
- Choose, if you can, vacation jobs that help you acquire career-useful skills or experiences (e.g. working with young people if your aim is teaching).
- After every job, note in your Personal Record the things you learned from doing it (e.g. team-work, cash-handling, punctuality, dealing with the public).
- Keep up any foreign language skills you have – holidays, the Internet, foreign newspapers and evening classes are all good for this.

- Employers want potential leaders. What have you led or organised recently? How could you develop your organisational talents?
- Organising some events for your department's geographical society might be a fun way of proving your skills as an organiser.
- What career skills or experiences do you think you can reasonably hope to acquire before the end of *this* academic year? What more could you do by the end of *next* year?
- Travel as widely as your budget will allow.
- Planning to travel the world this summer? Great! How can your travel be made to be of deeper career benefit?
- Enjoy yourself; believe in yourself.

GETTING A JOB

- Use the full resources of your careers service; there is a lot it can do for you and not just in your final year.
- List the features of the type of job you would ideally like. Re-read and amend this at the end of each year as your ideas and priorities change. Then you will know what you are aiming for and how to slant your geography degree to achieving it.
- Self-assessment is a useful way of tracking your growing range of skills, qualifications and experiences. Try using Appendices A and B to monitor your progress.
- About a quarter of geographers who graduate in the UK go on to further study. Should you be among them? Find out early from your careers service what courses and additional qualifications your chosen career might require.
- Get someone to read your curriculum vitae/résumé (you can return the favour by looking over his/hers) to eradicate any spelling mistakes and show you how others react to your style of writing (always hard to predict).
- Networking is a key skill in the workplace. It is never too soon to start networking.
- Employers want more than people with specific skills who can do the job. They want reasonably congenial colleagues. Are you such a person? Can you prove it?
- If you are not going to be among the 4–5 per cent of geography graduates who get a First, how are you going to stand out as being above-averagely employable? Can your extra-curricular activities help?

- There will be many opportunities for further study and research after graduation.
- How might you convince an employer that you are a *flexible* potential employee?
- Your chosen career area may not be obviously 'geographical'. How might you convince a recruiter that your geography degree is really relevant?

AND FINALLY . . .

- You are not, and never will be a failure. But everyone can do better.
- If the stress builds up, look at Section 2.13 for some tips.
- It's your degree and your education, not the university's.
- The world is getting smaller and the global economy needs geographers.
- 'Geography is about places; everything is somewhere; so geography is about everything.' True or False?
- Your learning will not stop when you leave university.
- Education is what is left after you have forgotten all the facts you learned.
- Need further help? Try the 'Further Reading' section at the end of this book

GOOD LUCK AND ENJOY GEOGRAPHY@UNIVERSITY!

APPENDIX A

SELF-ASSESSMENT OF SKILLS

You might find it helpful to measure your progress, in terms of your growing list of skills and experiences, and so highlighting the strengths of your employment profile. We suggest that you assess where you have got to periodically throughout your geography degree.

To record your annual progress (each column of the table), either enter (as appropriate) a number or put one tick against an item (meaning 'can do that') or two ticks ('really quite good at that'). Watch your profile improve as the years of your degree course go by. Of course, at the 'start of degree' stage most students will have only a few ticks.

	Start of Degree	End of Year 1	End of Year 2	End of Final Year
For the average lecture are your lecture notes one paragraph, one page or five pages? (two pages is about the best length)				
Can you write a fairly good essay in a week?				
How many computer packages can you use – at a basic level? – reasonably? – well?				
Can you word-process an essay?				
Can you send an e-mail message with an attachment?				
How many group projects have you worked on (at school or elsewhere)?				
How many talks or oral presentations have you given?				
Have you completed a major project or dissertation?				
How many foreign languages can you cope with – at a basic level? – reasonably? – well?				
Have you identified one or more possible career areas?				
How many events/projects/ groups of people have you successfully organised in the last 12 months?				

List here the practical field or research skills which you have learned and practised.

What other skills do you have? (specify them)

APPENDIX B

YOUR PERSONAL RECORD

Reflecting on your progress is time well spent. These pages, deliberately left largely blank, are designed to encourage you to keep a Personal Record of your skills, experiences and aims as they develop during the course of your geography degree. You can photocopy these pages and add to them after each year of your degree. You may find it helpful to record your progress under the headings we suggest below, but use any other headings you find useful. As your skills broaden and your ideas change, note the details here. This record can then form the basis for compiling your curriculum vitae or résumé which will accompany your letters of application for jobs in future years.

You are likely to have three types of skill and competencies:

- those skills you would like to claim to have but will find it hard to prove you have e.g. being hard working and reliable;
- those skills you claim to have and can provide some evidence for
  e.g. good ICT skills – 'I created my own website';
  can drive – ' have a full clean driving licence';
  good team member – 'I was part of a team of students who produced. . . etc.';
- those skills which are formally accredited
  e.g. doing research well – 'I got 75% for my dissertation';
  can speak French – 'I got an A grade in an examination'.

It will help convince potential employers if you can get some evidence or official accreditation or confirmation for as many of your skills and competencies as you can. Fortunately the wide range of teaching methods in geography courses should help you do this.

Your geography courses (what topics do you know about?)

Your range of communication skills:

– types of writing (e.g. essays, posters, field notebooks, projects);

– types of public speaking (e.g. brief tutorial talk, group presentation to class, individual presentation to class);

– created Website.

Your computing and ICT skills

Your statistical and graphical skills

Your research/project/dissertation skills

Your foreign-language skills and international/travel experience

Your organisational and group-work abilities

Your contacts with the world of work (your jobs, contacts, appreciation of what it means to be a successful employee)

Your social interests

Your personal strengths

What is important to you personally in your life and future career?

an excellent example of cause and effect.
nd large populations the effect. Looking back
lations increased relatively slowly for millions of
population only became alarming in the last
age of technology.

ples supporting how technology makes large
ical science and technology has allowed many
e irradicated, and consequently life expectency
s increased more in the last hundred years than
y. Technology has made it possible to convert
it was previously unsuitable. Large populations
endant on technology. In the developed world
le) is an essential part of the way we live, and
s we know it would be difficult – if not impos-
and less time is spent on basic tasks such as
ining and so on. Technology has been used to
. More and more time is devoted to leisure
t of sitting at home and watching one particular
tertainment and enlightenment: the television

iat large populations have made technology
our planet is of finite size. The supply of fossil
erves of metal ores are limited; the ability of
he toxic products of some forms of technology
echnology itself has limits. There comes a point
nology starts to produce ireversible changes in
in the ozone layer show this.
n recognised that a finite planet can support only
nly recently been widely accepted that our uses
ty can take place only on a limited scale. Despite
pulation of our planet, there has not been any
diness to accept the implications of the steps
n globally. Similarly, it seems difficult for people
orms of technology which are becoming life-

APPENDIX C

ESSAYS – THE GOOD AND THE BAD

Here are two essays on the same topic. Neither is without merit but both have clear faults which you should try to avoid in your essays. Can you spot the weaknesses of each essay? To orient you, both essays were written by first-year university students. Both are short essays of about 750 words, not the 2,500-word essays which are more commonly required. The students were told not to give references for their reading, so this is missing from both essays.

The things to look for in an essay are these.

1) Does the essay answer the question set?
2) Is it written in clear English with correct grammar and spelling?
3) Is there a clear argument in the essay or is it a jumble?
4) Does the essay have evidence of reading about the topic to discover more about it?
5) Does the essay try to get above the facts and deal with bigger issues relevant to the topic?

As you read the two essays mark them by these five criteria and see which comes out better on each.

ESSAY I: 'TECHNOLOGY MADE LARGE POPULATIONS POSSIBLE; LARGE POPULATIONS MAKE TECHNOLOGY INDISPENSABLE.' DISCUSS.

The population of every country in the world has increased in the last 150 years, and in some countries the rate of increase has been much greater than in Britain. When populations increase more food has to be produced to keep these people nourished. More houses have to be built, with increased water supply and facilities for sewage disposal. More schools and hospitals are needed. Forests are cleared and burnt to give fertile land to grow crops, but if the increase in food production and provision of housing and other facilities does not take place at the same rate as the increase in population, then the standard of living falls.

The life expectancy up to about 150 years ago rose slowly from an average of 20 years for early man to the 50s in the nineteenth century. Today's life expectancy is 70–80 years and is therefore a population increase of 50%. Infant and child mortality has until now, been high – in 1866 only 40–45% of those aged 0–15 years were expected to survive. Diseases such as cholera, typhoid, tuberculosis and diphtheria took their toll, mainly among the poor living in slums, as Marx noted in Manchester. Improvements in the standard of hygiene and sanitation reduced this rate so that the survival rate of 0–15 year olds increased to between 62% and 75% in 1961 in Britain, but it is lower than this in many parts of the world today.

The world population in 3000 BC was about 100 million. This slowly increased to about 350 million in 100 BC. In the Middle Ages the population was steady at around 350–500 million until the Black Death caused a drop to 250 million people. The population then increased rapidly to around 1000 million in the 1800s and then very rapidly to nearly 4000 million in the early 1980s. The projected rise according to United Nations figures is to around 6500 million people by the year 2020.

Darwin stated in his book *The Origin of Species* that 'there is no exception to the rule that every organic being naturally increases at so high a rate that if not destroyed, the earth would soon be covered by the progeny of a single pair'. He also recognised that there are many 'checks to increase' which limit the size of populations. For humans, the checks have included diseases and epidemics as well as famine over which until recent times, man has had very little control, and wars over which he does. Even now, his control is far from complete where diseases are concerned. The degree to which the incidence of fatal diseases such as smallpox, cholera and tuberculosis has been reduced since 1955, has resulted in an enormous increase in population throughout the tropics. With more mouths to feed, the threat of starvation has increased. Starving people are not as well able to work and increase food production as healthy, well-fed people.

The rise in population is exponential and one wonders what the next 'check to increase' can be. Already, shortages of food and water (e.g. in

the S
man
woul
stanc

H
com
appr
little
men
phys
or jo
of lif
othe
great

Th
resou
has t

·

·

**ESSAY 2: 'TECHNOLOGY
POSSIBLE; LARGE POPUL
INDISPENSABLE.' DISCUS**

At first sight this seem
Technology is the cause
over human history, pop
years, and the growth i
hundred years or so – th

There are many exa
populations possible. M
former killer diseases to
in the developed world h
in our evolutionary histo
to agricultural use land t
have indeed become de
transportation (for exam
without technology, life
sible. In the home, less
cooking, washing and cl
automate many of thes
activities – particularly th
form of technology for e
set.

So it would appear
indispensible. However,
fuels will run out; the re
our ecosystem to absorb
is limited. In other words,
where greater uses of tec
our ecosystem. The hole
While it has always bee
a finite population, it has
of some forms of technolo
the logic of limiting the p
evidence of any real rea
necessary to limit populat
to agree to curb some
threatening.

All is not bad news, however. There are some sources of technology which do not have any deliterious effects on our environment. For example, hydroelectricity, solar energy, wind power, tidal power and wave power are all 'free' forms of energy which do not cause toxins to be released into our ecosystem, and potentially allow us to continue to be dependant on various electrical technologies. Other renewable resources can allow us to continue using paper, some liquid fuels, and medecines. Often, the cost of using renewable resources is greater that that of consuming the finite reserves of our planet, but if large populations are to continue to depend on technology, that cost – sooner or later – will have to be accepted.

So far we have argued that technology causes (or at least permits) an increase in population without the impoverishment of everyone. Equally one could argue that, as population increases, the need for new ways of living creates a climate for inventions to be sought, new products to be worked up into saleable goods, and new technologies to be diffussed globally. Population pressure induces its new technologies for survival (eg food production) or an enhanced quality of life (eg better transport).

In conclusion, it is certainly true that technology played a major part in allowing large populations to develop. To sustain such populations, technology will indeed be indispensible – but only if the wisdom of choosing *appropriate* technologies is brought to bear on preserving our ecosystem. The relationship between technology and population is therefor variable, depending on which techologies are chosen and how they are used. Equally technology advances as much as a product of population growth as its cause.

- What are the strengths and weaknesses of this short essay?

- What is your suggested mark (%)?

REVIEW

You have probably come to the conclusion that Essay 1 has plenty of facts (though no formal referencing of sources) and the author has clearly been reading around the subject, which is good. However, it lacks a structure to the extent that it hardly answers the question. Despite good grammar, spelling and English it would get a poor mark – probably well under 50%. Essay 2 has too many spelling mistakes and it is almost completely devoid of any facts or signs of reading. Its statements, even if correct, are backed up by no evidence. It does, however, have some rather interesting, even clever ideas about the relationships between population and technology.

A good essay would combine the best features of both essays – a clear argument backed up by referenced factual material as evidence, and some high-level ideas about the bigger issues. That would be a good answer to the question set.

APPENDIX D

WEB RESOURCES FOR GEOGRAPHERS

The web resources which the undergraduate geographer can use to help with his/her studies are vast and rapidly expanding. The sites are also changing in terms of their web addresses and structure. So the sites listed below are those available early in 2002. You have to expect that these sites will become dated or inaccessible within a few years – that is the nature of the web.

In Section 4.4 we gave some advice about treating all websites with caution. That advice applies to those listed below. We may think them interesting and potentially useful, but we cannot foresee the uses you will wish to make of them and they may be more helpful for some purposes than for others. *The Internet Detective* site gives some advice on how to evaluate the quality of websites in general: http://sosig.ac.uk/desire/internet-detective.html

GENERAL REFERENCE

1) A good general site is this American one:
   http://www.colorado.edu/geography/virtdept/resources/contents.htm

2) The CTI Centre's site at Leicester is also worth a look because of its clear thematic organisation, though the site may not be updated as often as some others:
   http://www.geog.le.ac.uk/cti/info.html

3) The BBCi site is an eclectic collection that could prove useful in some cases:
   http://www.bbc.co.uk/categories

4) For basic factual material, interesting sites are the Guinness Book of Records at:
http://www.guinnessrecords.com/

and the Yahoo geography website at:
http://education.yahoo.com/reference/factbook/

5) The 'Dictionary of Units' site tells you about units of measurement and conversion factors between units:
http://www.ex.ac.uk/cimt/dictunit/dictunit.htm

6) The 'xrefer' site is the equivalent of an online dictionary – good for looking up the definitions of things in a wide variety of areas:
http://w1.xrefer.com/allfreebooks.jsp

MAPS

1) For UK maps try the Ordnance Survey's Get-A-Map service at:
http://www.ordsvy.gov.uk/

which has links to historical maps, or the Multimap site at:
http://www.multimap.co.uk

2) For street maps try:
http://www.streetmap.co.uk
which has links to aerial photographs.

3) A big library of downloadable maps of much of the world is held at the University of Texas:
http://www.lib.utexas.edu/maps/index.html

4) Digimap allows you to custom-build your own maps but you need to register with the service and this is easier if you are a member of a university which has a site licence:
http://digimap.edina.ac.uk

5) For non-UK maps try the National Geographical Society's site at:
http://www.nationalgeographic.com/maps/index.html

or MapQuest at:
http://www.mapquest.com/

6) Historical UK maps are available from:
http://www.old-maps.co.uk/

7) The more adventurous might like to try their hand at the online creation of their own maps at this site:
http://www.aquarius.geomar.de/omc/

REMOTE SENSING

1) European weather satellite images are available from the Dundee Satellite Receiving Station at:
http://www.sat.dundee.ac.uk

2) Aerial photographs of the UK can be viewed from the GetMapping site but cannot really be downloaded unless you buy them:
http://www.getmapping.com

3) A good instructional site on remote sensing is at:
http://www.nln.met.ed.ac.uk/

4) A good starting point for RS imagery and RS links is:
http://www.mimas.ac.uk/rs/

5) This NASA site has a wide range of material that could be useful in several areas of geography:
http://earthobservatory.nasa.gov/Search/sitemap.html

STATISTICS – METHODS AND DATA

1) Two good online statistics courses to brush up on methods are available at:
http://davidmlane.com/hyperstat/index.html
http://www.statsoftinc.com/textbook/stathome.html

2) The best source of UK statistics is the site of the Office for National Statistics at:
http://www.statistics.gov.uk

3) You can also find additional material on many of the websites of individual government departments. All national and local government

departments' and agencies' websites are listed alphabetically and by topic or keyword at:
http://www.ukonline.gov.uk/

4) The (UK) Social Sciences Information Gateway site also gives access to statistical material on many topics:
http://www.sosig.ac.uk

NEWS AND CURRENT AFFAIRS

1) Probably the best site is that of the BBC which is extensive, well indexed, easy to search and has as good a historical and global coverage as any online source:
http://news.bbc.co.uk

2) Also worth looking at are the CNN site:
http://www.cnn.com/

and the Reuters site at:
http://www.reuters.com/

COUNTRIES

1) BBC World Service gives up-to-date news and country background profiles for the whole world at:
http://www.bbc.co.uk/worldservice/

2) The CIA also provides country profiles at:
http://www.odci.gov/cia/publications/factbook/index.html

3) The United Nations provides statistical data on its member states at:
http://www.un.org/Pubs/CyberSchoolBus/infonation/e_infonation.htm

4) The Yahoo site also gives current affairs information as well as socio-economic material on individual countries:
http://dir.yahoo.com/regional/countries/

PHYSICAL GEOGRAPHY AND EARTH/ENVIRONMENTAL SCIENCES

1) *Fundamentals of Physical Geography* is a Canadian site which acts as an online physical geography course:
http://www.geog.ouc.bc.ca/physgeog/home.html

2) More physical geography resources, listed by topic, are available at:
http://personal.cmich.edu/~franc1m/homepage.htm

3) Links to national meteorological services across the world are provided by the UK Met. Office's website:
http://www.metoffice.gov.uk/corporate/links/nms.html

4) Although entitled Weatherworks, this site is broader in its coverage of useful atmospheric and environmental links:
http://www.weatherworks.com/links.html

5) Tidal information from the Proudman Oceanographic Institute is available at:
http://www.pol.ac.uk

6) Weather-related information on the Tiempo site in Norwich is available at:
http://www.cru.uea.ac.uk/tiempo/site.htm

HUMAN GEOGRAPHY

1) Human geography websites are rarer than physical geography ones. Here is an American one:
http://dir.yahoo.com/Science/Geography/Human_Geography/

2) And this is a British one which also gives links to some geographical journals:
http://www.sosig.ac.uk/roads/subject-listing/World-cat/geog.html

3) For development studies, the following are useful sites, especially for their statistical material:
http://www.adb.org (Asian Development Bank)
http://www.ebrd.com (the European Bank for Reconstruction and Development – good for environmental and other issues in Eastern Europe and the Former Soviet Union)

http://www.imf.org (the International Monetary Fund – fairly specialised and macro-economic, but gives rapid, free access to World Economic Outlook Reports)

http://www.oecd.org (the site of the Organisation for Economic Co-operation and Development which covers a wide range of geographical topics)

http://www.sei.se (the Stockholm Environmental Institute)

http://www.worldbank.org (The World Bank site has huge amounts of useful statistics and other material)

4) Statistical and particularly demographic data for most countries of the world are available from this US Census Bureau site:
http://www.census.gov/ipc/www/idbnew.html

GEOGRAPHY STUDY SKILLS

1) It is worth looking at the website associated with Pauline Kneale's book, *Study Skills for Geography Students: a Practical Guide* (Arnold, 1999):
http://www.geog.leeds.ac.uk/staff/p.kneale/skillbook.html

2) And the website of the Geography for the New Undergraduate project at Liverpool Hope University:
http://www.livhope.ac.uk/gnu/

CAREERS

1) Ideas on occupations and how they would suit you can be obtained from this site, particularly the Quick Match option:
http://prospects.csu.ac.uk/student/cidd/main.stm

2) Advice on job hunting via the Internet is given at:
http://careers-main.lancs.ac.uk/findinet.htm

3) Earthworks is an online list of job vacancies for geographers and environmental scientists:
http://www.earthworks-jobs.com/

4) Postgraduate course and research opportunities can be explored at:
http://www.prospects.csu.ac.uk/pg/

5) Access to university online prospectuses for further information on particular courses can be obtained from:
http://www.scit.wlv.ac.uk/ukinfo/

6) An excellent Australian careers site (and useful not only to those in Australia) is at:
http://www.iag.org.au/careers.htm

APPENDIX E

GOOD EXAMINATION ANSWERS

First, one must stress that there is no one model of a good examination answer. Much depends on the types of question set – and they do vary. There are always different routes to a good examination mark, unless it is a very tightly controlled multiple-choice type of examination. However, that does not help you. How might you use the general advice on essay writing given in Section 5.4 to construct an answer? Here is an example.

Suppose that you have chosen to answer this examination question: **Describe and explain the changes in UK forestry policy since 1919**. How might you structure and write an answer in 45 or 60 minutes?

First, you need to analyse the elements in the question's title. There are two tasks – to *describe* and to *explain*; you need to do both. The focus is on *policy* and *changes in policy*, not on, say, the environmental effects of forestry. There are clear restrictions on the type of material you can use:

- area = UK;
- period = since 1919;
- field = forestry.

If you do not know about these issues you cannot substitute non-British, historical or agricultural material.

A possible and fairly 'safe' structure for an answer would be chronological. Define a series of periods from 1919 to the present and for each describe the policy and policy aims operated then and explain why they were used, what forces built up to alter the policy in a later period and the way the change to the new policy took place. For example:

Paragraph 1
Why was the Forestry Commission set up in its then form in 1919? What was new about this compared with pre-1914 policy?

Paragraph 2
What was policy trying to do in the inter-war period – public spending cuts, fear of a new war, Great Depression, self-sufficiency?

Paragraph 3
The post-war period might be divided into three overlapping periods such as:

- the 1950s and 1960s when the policy aims were **a**, **b** and **c**;
- the 1960s and 1970s when the aims were **c** (continuing), **d** and **e**;
- and the 1980s and 1990s when the most recent aims to be introduced were **g** and **h**.

Try to link the changes in forestry to wider changes in the UK economy or politics (e.g. combating the Depression, winning the Second World War, Thatcherism).

Leave room in your 45–60 minute answer for the most recent events (e.g. environmentalism, privatisation, community forests) by not spending too much time on the earlier periods.

Leave room also for a final paragraph which will summarise the main trends, and briefly note the major continuities over the whole period since 1919 and the major shifts since then. Is there any overall model that would characterise British forestry policy in this period? So, if Brown put forward a model of a transition from an industrial forestry model to a post-industrial one, make sure you include this 'higher-level' approach to the subject. If there have been controversies between Jones and Smith on how to explain inter-war forestry policy and between the Forestry Commission and the Treasury on the economic justification for UK forestry, cite those authors and include their ideas. Referencing the literature is always important in examination answers.

The answer should be grounded in real events so make reference to actual schemes, places, authors and forestry statistics (e.g. timber self-sufficiency ratio, percentage of UK afforested) to illustrate your points.

Here are other possible examination questions:

Describe and account for the principles of town planning used in the UK and USA since 1960.

Comments

1) You have to do what the question tells you. So, it would be wrong just 'to describe' and not also 'to account for'.

2) 'The principles' – it would be wrong to focus on one principle and ignore the others. One principle may be dominant and get more attention, but this should not be to the exclusion of the other principles.

3) 'In the UK and USA' implies substantial though not necessarily equal coverage of both countries. References to other countries should be limited to key comparisons or statements of where UK/US planning ideas came from or were subsequently adopted.

4) 'Since 1960' means the whole period, not one decade. If one sub-period was particularly influential then it can get more attention but not to the exclusion of the rest of the period. Material prior to 1960 should be used very sparingly, perhaps to sketch the origins of post-1960 activity.

Explain the reform of the Common Agricultural Policy since 1980.

Comments

1) It would be wrong to spend so much time describing the CAP and its effects and weaknesses that one is left with little time to deal with its reform (the focus of the question). 'The reform' implies some coverage of what is being reformed (the current structure) and why reform is needed (the problems with the CAP), but there must be good coverage of what reforms are in place and their effects and of what reforms are proposed.

2) The question could be answered from an economic perspective (the effects in terms of incomes, costs and production on farmers and consumers in different places) or from a political perspective (the power blocks lobbying for and against particular reforms). A good answer would include both perspectives on reform.

3) 'The reform' is a singular noun but that does not mean that the CAP has only one set of critics, one problem and one possible direction of reform. A good answer might note how the criticisms, arguments and types of reform have varied (i) over time and (ii) among the different countries and interest groups. Good examination answers should show that you have understood the complexities of the topic, can describe them clearly and can give your reasoned view of the best way to see the issue.

APPENDIX F

MARKING CRITERIA FOR DISSERTATIONS

Each university will have slightly different ways of describing the qualities required in a geography dissertation to be awarded particular marks, but what is given below is a good guide to what markers are looking for. In other words, if you can ensure that your dissertation has the qualities associated with the higher mark ranges, then it should get a good mark.

That means thinking about the qualities of a good dissertation as soon as you start it, so you can build in methods, theories and analyses that will be applauded by your examiners.

90–100%
An exceptional dissertation, excellent in every respect. A highly appropriate, intellectually demanding and original topic with extremely well-defined aims identified within a very well understood conceptual framework based on extensive, if not exhaustive, understanding of the literature. Fieldwork or other sources are used extensively and extremely effectively. Methodology and data collection are thorough, comprehensive and innovative. High-quality results, interpretations and discussion demonstrate an outstanding ability to analyse, synthesise and evaluate. The dissertation is very well organised, sharply focused and stylishly written. Presentation is flawless, possibly of publishable quality.

80–89%
An outstanding dissertation, excellent in almost all respects. A highly appropriate and intellectually demanding if not original topic with well defined aims identified within a well-understood conceptual framework based on extensive understanding of the literature. Fieldwork or other sources are used extensively and very effectively. Methodology and data

collection are thorough, comprehensive and may be innovative. High-quality results, interpretations and discussion that show excellent cognitive skills. The dissertation is very well organised and sharply focused with a high standard of presentation.

70–79%

An excellent dissertation in most respects. An appropriate and intellectually demanding if not necessarily original topic with well-defined aims, within a well-understood conceptual framework based on extensive background reading. Fieldwork or other sources are used extensively and effectively. Methodology and data collection are thorough. Results are detailed and accurate. Interpretation, analysis and discussion show very good cognitive skills. Conclusions are substantial. The dissertation is well organised with a high standard of presentation.

65–69%

A very good dissertation on a well-chosen topic with clearly stated aims identified within an explicit conceptual framework based on significant background reading. Extensive use of original data, though perhaps not to full effect. Well-chosen methodology but possibly with minor design flaws. Data collection at least to accepted minimum levels (e.g. appropriate numbers of questionnaires, interviews). Results are detailed and mostly accurate. Interpretation, analysis and discussion show satisfactory cognitive skills. Conclusions are sound and clearly related to the aims. May be let down by some errors or omissions. The dissertation is well presented.

60–64%

A good dissertation with clearly stated aims identified within an explicit conceptual framework based on a satisfactory level of background reading. A reasonable amount of original data have been collected but data collection may have limitations and methodology may have design flaws. Results are mostly accurate but may contain errors and omissions. Interpretation, analysis and discussion may exhibit weaknesses, especially in evaluation and synthesis. Conclusions are sound and there is at least an attempt to relate them back to the aims. The dissertation may contain weaknesses in organisation but is generally well presented.

55–59%

A competent dissertation with specified aims within a recognisable conceptual framework possibly based on sound but limited background reading. The topic may be solid but uninspiring. Data collection may be insufficient to get the best out of the project and may be over-reliant on secondary sources. Though execution may be adequate, there may be only basic justification of a flawed methodology. If data collection reaches minimum levels it may be flawed in other ways. Results or analyses may contain errors as well as omissions. Interpretation and discussion may be poorly developed. Tends to be descriptive rather than analytical; may contain superfluous or irrelevant material. Conclusions may be sound but unfocused. Weaknesses may be evident in organisation or presentation.

50–54%

An adequate dissertation with aims, possibly poorly specified, within a recognisable conceptual framework based on limited background reading. The topic may have been poorly chosen. Original data collection may be insufficient with an over-reliance on secondary material. Though execution may be adequate, methodology may be weak and insufficiently justified. If data collection reaches minimum levels it may be flawed in other ways. Results and analyses may contain errors as well as omissions. Interpretation and discussion are poorly developed. The dissertation may not distinguish relevant material from the irrelevant and superfluous. Conclusions may be repetitive or unfocused. Weaknesses are likely in organisation and presentation.

45–49%

A deficient dissertation with poorly specified aims and/or inadequate conceptual framework based on little background reading. The topic may have been poorly chosen. Limited collection of original data and heavy reliance on secondary material are likely. Methodology may be unexplained and data collection may fall below minimum levels. Results usually contain errors and omissions. Attempted analyses may be inappropriate. Interpretation and discussion may be limited with irrelevant material. Conclusions may be repetitive, unfocused and incomplete. The dissertation may be poorly organised with serious flaws in presentation. Strengths tend to be mainly those of effort and persistence; though the content has some merit, little of the possible potential has been realised.

40–44%

A weak dissertation with poorly specified aims and/or inadequate conceptual framework based on minimal background reading. Significant weaknesses are likely in planning and implementation. Conceptual content may be minimal. Methodology may be unexplained and little original data collected. Results contain significant errors and omissions. Analysis may be absent. Significant deficiencies are evident in interpretation, discussion, conclusions in terms of focus, expression, length, completeness and organisation. Presentation may be barely acceptable.

35–39%

Not an honours standard dissertation; exhibits serious deficiencies such as aims poorly defined or lacking, little or no conceptual framework, methodology inappropriate or misunderstood, data collection inadequate or non-existent, poor description of results, lacking analysis, wrong interpretations, limited discussion, superficial conclusions and barely acceptable presentation.

20–35%

A very poor dissertation, showing few signs of having been taken seriously. There is only a very limited attempt to introduce the topic, describe methods, present and discuss results, and come to a conclusion.

0–19%

Exceptionally poor dissertation, showing no signs of having been taken seriously.

APPENDIX G

GIVING A SEMINAR PRESENTATION –
SOME PRACTICAL ADVICE

Giving an oral presentation can be a daunting experience, particularly if you are inexperienced, but it is a skill which can be readily acquired and whose long-term value to you will be considerable.

If you are sharing the presentation with others, you need to divide the time and the topic between you so that each person knows his/her role. Time is of the essence. If you have to give a 30-minute presentation, that is how long it should be. You will need to practise the presentation to ensure it is the correct length; tape-recording your practice sessions can be enlightening. The more you have rehearsed, the easier it will be on the day.

You will need a script of what to say which will also remind you when to show slides or overhead transparencies. You should try to avoid reading the script; refer to the script when you need to, in between talking to the audience. Eye contact (looking at the audience) is the ideal to aim for.

You need to be audible (neither too loud nor too quiet) and to speak at a comfortable pace for an audience who will be trying to take notes. So you must not talk too quickly, or so slowly that you are tedious. Excessive speed of delivery is the commonest fault among those new to public speaking. An erect posture rather than slouching usually improves your audibility.

An oral presentation, even more than a written essay, needs a very clear, early statement of the purpose and structure of the talk: an overhead projector may help here. A handout showing graphs, data, key facts, references or illustrations can also add to the presentation's effectiveness. Try to avoid a monotonous tone and make the topic sound reasonably interesting.

Finally, a useful short-cut to becoming a more proficient speaker is to try to copy the good points of your better lecturers and avoid the worst features of the poorer ones.

OVERHEAD PROJECTION

Overhead transparencies (OHTs or 'foils') can be a useful device for illustrating a presentation. You can show illustrations (e.g. graphs or maps, sometimes even photographs), give key facts (e.g. names or statistics) or set out the structure of the talk.

Being a visual medium, an overhead transparency's effectiveness depends on its visual clarity. Therefore the text must be clear (rather than faint) and large enough to be legible even at the back of the lecture room you will be using (check this out). Colour can add to the effectiveness of an OHT if it highlights the key points, but don't overdo the colour. If the OHT comprises text, neat lettering by hand can be adequate. Output from a laser printer at an appropriate size (say, 20–40 point) is excellent. If the illustration is taken from another publication it is wise to.

(i)   use an enlarging/reducing photocopier to get the illustration to the right size;
(ii) cut off extraneous material (for example, the surrounding text) before making the OHT;
(iii) check that the exposure level is correct (neither too dark nor pale).

If you have several OHTs in your presentation, it is wise to number them in sequence (in case you drop them) and write their numbers in the margin of your script as a cue to their display. It can be helpful to separate the sheets with a piece of white paper.

Material can be photocopied directly on to an OHT, but you must use the correct type of acetate sheet. There are several types of acetate sheet for overhead projection:

* 'write-on film' which is for your own writing in pen and must **not** be used in a photocopier (it fries inside the machine);
* OHT photocopier film (which can be used in a photocopier).

The two types can look similar so check that you are using the correct one.

Before the presentation, check the overhead projector you will be using. Ensure that the image is of the right size on the screen and is in focus. When showing an OHT ensure that you do not block the image on the screen by standing directly in front of the projector.

HANDOUT

A handout is a useful way of giving the audience the following types of material:

* a guide to the lecture's structure;
* references;
* diagrams;
* factual material (e.g. definitions, dates, formulae, lists, case studies).

For ease of use, staple the sheets together, number the pages, and refer to the appropriate part of the handout during the talk.

SLIDES

Check the slide projector and how to work it well before the talk. Ensure that you load the carousel before the talk and run through the slides to ensure that they are all the right way up and right way round.

Check the slides are visible even from the back of the room. Ensure that the level of ambient lighting is neither too high (cannot see the slides) nor too low (audience falls asleep).

POWERPOINT

A PowerPoint display lets you combine in one technology the visual images of the photographic slide and the website with the text and graphics of the overhead transparency. Once you have got access to the PowerPoint software, you need to ensure that you do not spend too much time on the many stylistic options of that package (the colours, visual effects, etc.) and focus on the content of the display.

It is essential that you practise with PowerPoint on the equipment in the room you will be using. PowerPoint, particularly when used through a network system, can be unreliable, so it is a wise precaution to have a spare set of your images as overhead transparencies, so you can revert to the older technology if the newer one lets you down.

A useful reference for more sound practical advice on giving a presentation is:

Hay, I. (1994) 'Notes of guidance for prospective speakers', *Journal of Geography in Higher Education*, 18(1): 57–65.

FURTHER READING

The items marked with an asterisk (*) are perhaps the best in each section and the ones to go for if your time for further reading is limited.

GEOGRAPHY STUDY SKILLS

* Kneale, P. (1999) *Study Skills for Geography Students: a Practical Guide*. London: Arnold.
Lindsay, J. M. (1997) *Techniques in Human Geography*. London: Routledge.

HOW TO BE A STUDENT AND GET A DEGREE

Barnes, R. (1995) *Successful Study for Degrees*. London: Routledge.
Dweck, C. (2000) *Self-Theories: Their Role in Motivation, Personality, and Development*. Philadelphia, PA: Taylor and Francis.
Honey, P. and Mumford, A. (1982) *The Manual of Learning Styles*. Maidenhead: Peter Honey.
Marshall, P. (1995) *How to Study & Learn: Your Practical Guide to Effective Study Skills*. Plymouth: How To Books.
Miller, C. M. L. and Parlett, M. (1974) *Up to the Mark: a Study of the Examination Game*. Monograph 21, London: Society for Research in Higher Education.
* Northedge, A. (1995) *The Good Study Guide*. Milton Keynes: The Open University.
* Rowntree, D. (1998) *Learn How to Study: a Realistic Approach*. London: Warner Books.
Tolmie, P. (ed.) (1998) *How I Got My First Class Degree*. Lancaster: Unit for Innovation in Higher Education, Lancaster University.

WHAT GEOGRAPHY IS ABOUT

Allen, J. and Massey, D. (1995) *Geographical World*. Oxford: Oxford University Press.
Gould, P. (1985) *The Geographer at Work*. London: Routledge.

Haggett, P. (1990) *The Geographer's Art.* Oxford: Basil Blackwell.

Massey, D. and Allen, J. (eds) (1984) *Geography Matters! A Reader.* Cambridge: Cambridge University Press.

Quality Assurance Agency (2000) *Geography.* Gloucester: Quality Assurance Agency.

Also at http://www.qaa.ac.uk/crntwork/benchmark/geography.pdf

* Rogers, A., Viles, H. and Goudie, A. (1992) *The Student's Companion to Geography.* Oxford: Blackwell.

EXAMINATIONS AND ESSAYS

Barrass, R. (1995) *Students Must Write.* London: Routledge.

Becker, H. (1986) *Writing for Social Scientists.* Chicago: University of Chicago Press.

* Creme, P. and Lea, M. (1997) *Writing at University: a Guide for Students.* Buckingham: Open University Press.

Fitzgerald, M. (1994) 'Why write essays?', *Journal of Geography in Higher Education,* 18(3): 379–84.

Hay, I. (1995) 'Writing a review', *Journal of Geography in Higher Education,* 19(3): 357–63.

Hay, I. (1996a) 'Examinations I', *Journal of Geography in Higher Education,* 20(1): 137–42.

Hay, I. (1996b) 'Examinations II', *Journal of Geography in Higher Education,* 20(2): 259–64.

Hay, I. (1997) *Communicating in Geography and Environmental Sciences.* Oxford: Oxford University Press.

Mills, C. (1994) 'Acknowledging sources in written assignments', *Journal of Geography in Higher Education,*18(2): 263–8.

Sussams, J. E. (1998) *How to Write Effective Reports.* London: Gower.

HOW TO PRODUCE A DISSERTATION

Bell, J. (1993) *Doing Your Research Project.* Buckingham: Open University Press.

Burkill, S. and Burley, J. (1996) 'Getting started on a geography dissertation', *Journal of Geography in Higher Education,* 20(3): 431–8.

* Flowerdew, R. and Martin, D. (1997) *Methods in Human Geography: a Guide for Students Doing a Research Project.* Harlow: Longman.

Kitchin, R. and Tate, N. (2000) *Conducting Research in Human Geography: Theory, Methodology and Practice.* Harlow: Prentice Hall.

Parsons, A. J. and Knight, P. (1995) *How To Do Your Dissertation in Geography and Related Disciplines.* London: Chapman and Hall.

Turabian, K. (1987) *A Manual for Writers of Term Papers, Theses and Dissertations.* Chicago: University of Chicago Press.

OTHER ASPECTS OF GEOGRAPHY COURSES

Hay, I. (1994) 'Notes of guidance for prospective speakers', *Journal of Geography in Higher Education*, 18(1): 57–65.

* Hay, I. (1997) *Communicating in Geography and Environmental Sciences*. Oxford: Oxford University Press.

Lee, P. and Stuart, M. (1997) 'Making a video', *Journal of Geography in Higher Education*, 21(1): 127–34.

Lewis, S. and Mills, C. (1995) 'Field notebooks: a student's guide', *Journal of Geography in Higher Education*, 19(1): 111–14.

Vujakovic, P. (1995) 'Making posters', *Journal of Geography in Higher Education*, 19(2): 251–6.

Vujakovic, P., Livingstone, I. and Mills, C. (1994) 'Why work in groups?', *Journal of Geography in Higher Education*, 18(1): 124–7.

Young, C. (1998) 'Giving oral presentations', *Journal of Geography in Higher Education*, 22(2): 263–8.

APPLYING FOR JOBS

There are many books on this topic in high-street bookshops and there are national traditions in how to apply for jobs which need to be borne in mind.

Bates, T. and Bloch, S. (1997) *Employability: How To Get Your Career on the Right Track*. London: Kogan Page.

REFERENCES

Allen, J. and Massey, D. (1995) *Geographical World*. Oxford: Oxford University Press.

Barnes, R. (1995) *Successful Study for Degrees*. London: Routledge.

Barrass, R. (1995) *Students Must Write*. London: Routledge.

Bates, T. and Bloch, S. (1997) *Employability: How To Get Your Career on the Right Track*. London: Kogan Page.

Becker, H. (1986) *Writing for Social Scientists*. Chicago: University of Chicago Press.

Bell, J. (1993) *Doing Your Research Project*. Buckingham: Open University Press.

Burkill, S. and Burley, J. (1996) 'Getting started on a geography dissertation', *Journal of Geography in Higher Education*, 20(3): 431–8.

Creme, P. and Lea, M. (1997) *Writing at University: a Guide for Students*. Buckingham: Open University Press.

Dweck, C. (2000) *Self-Theories: Their Role in Motivation, Personality, and Development*. Philadelphia, PA: Taylor and Francis.

Fitzgerald, M. (1994) 'Why write essays?', *Journal of Geography in Higher Education*, 18(3): 379–84.

Flowerdew, R. and Martin, D. (1997) *Methods in Human Geography: a Guide for Students Doing a Research Project*. Harlow: Longman.

Gould, P. (1985) *The Geographer at Work*. London: Routledge.

Haggett, P. (1990) *The Geographer's Art*. Oxford: Basil Blackwell.

Hampson, L. (1994) *How's Your Dissertation Going?* Lancaster: Unit for Innovation in Higher Education, Lancaster University.

Hay, I. (1994) 'Notes of guidance for prospective speakers', *Journal of Geography in Higher Education*, 18(1): 57–65.

Hay, I. (1995) 'Writing a review', *Journal of Geography in Higher Education*, 19(3): 357–63.

Hay, I. (1996a) 'Examinations I', *Journal of Geography in Higher Education*, 20(1): 137–42.

Hay, I. (1996b) 'Examinations II', *Journal of Geography in Higher Education*, 20(2): 259–64.

Hay, I. (1997) *Communicating in Geography and Environmental Sciences*. Oxford: Oxford University Press.

Honey, P. and Mumford, A. (1982) *The Manual of Learning Styles*. Maidenhead: Peter Honey.

Kitchin, R. and Tate, N. (2000) *Conducting Research in Human Geography: Theory, Methodology and Practice*. Harlow: Prentice Hall.

Kneale, P. (1999) *Study Skills for Geography Students: a Practical Guide*. London: Arnold.

Lee, P. and Stuart, M. (1997) 'Making a video', *Journal of Geography in Higher Education*, 21(1): 127–34.

Lewis, S. and Mills, C. (1995) 'Field notebooks: a student's guide', *Journal of Geography in Higher Education*, 19(1): 111–14.

Lindsey, J. M. (1997) *Techniques in Human Geography*. London: Routledge.

Marshall, P. (1995) *How to Study & Learn: Your Practical Guide to Effective Study Skills*. Plymouth: How To Books.

Massey, D. and Allen, J. (eds) (1984) *Geography Matters! A Reader*. Cambridge: Cambridge University Press.

Miller, C. M. L. and Parlett, M. (1974) *Up to the Mark: a Study of the Examination Game*. Monograph 21, London: Society for Research in Higher Education.

Mills, C. (1994) 'Acknowledging sources in written assignments', *Journal of Geography in Higher Education*, 18(2): 263–8.

Northedge, A. (1995) *The Good Study Guide*. Buckingham: The Open University.

Parsons, A. J. and Knight, P. (1995) *How To Do Your Dissertation in Geography and Related Disciplines*. London: Chapman and Hall.

Phillips, E. and Pugh, D. (1994) *How To Get a PhD*. Buckingham: Open University Press.

Quality Assurance Agency (2000) *Geography*. Gloucester: Quality Assurance Agency.
Also at http://www.qaa.ac.uk/crntwork/benchmark/geography.pdf

Rogers, A., Viles, H. and Goudie, A. (1992) *The Student's Companion to Geography*. Oxford: Blackwell.

Rowntree, D. (1998) *Learn How to Study: a Realistic Approach*. London: Warner Books.

Sussams, J. E. (1998) *How to Write Effective Reports*. London: Gower.

Tolmie, P. (ed.) (1998) *How I Got My First Class Degree*. Lancaster: Unit for Innovation in Higher Education, Lancaster University.

Turabian, K. (1987) *A Manual for Writers of Term Papers, Theses and Dissertations*. Chicago: University of Chicago Press.

Vujakovic, P. (1995) 'Making posters', *Journal of Geography in Higher Education*, 19(2): 251–6.

Vujakovic, P., Livingstone, I. and Mills, C. (1994) 'Why work in groups?', *Journal of Geography in Higher Education*, 18(1): 124–7.

Young, C. (1998) 'Giving oral presentations', *Journal of Geography in Higher Education*, 22(2): 263–8.

INDEX